Graduate Texts in Mathematics 253

Graduate Texts in Mathematics

For other titles published in this series, go to
http://www.springer.com/series/136

Barbara D. MacCluer

Elementary Functional Analysis

 Springer

Barbara D. MacCluer

Department of Mathematics
University of Virginia
P.O.Box 400137
Charlottesville VA 22904-4137
USA
bdm3f@virginia.edu

ISBN: 978-1-4419-2753-8
DOI: 10.1007/978-0-387-85529-5

e-ISBN: 978-0-387-85529-5

Mathematics Subject Classification (2000): 46-01, 47-01

To Tom, Josh, and David

Preface

Functional analysis arose in the early twentieth century and gradually, conquering one stronghold after another, became a nearly universal mathematical doctrine, not merely a new area of mathematics, but a new mathematical world view. Its appearance was the inevitable consequence of the evolution of all of nineteenth-century mathematics, in particular classical analysis and mathematical physics. Its original basis was formed by Cantor's theory of sets and linear algebra. Its existence answered the question of how to state general principles of a broadly interpreted analysis in a way suitable for the most diverse situations.
A.M. Vershik ([45], p. 438).

This text evolved from the content of a one semester introductory course in functional analysis that I have taught a number of times since 1996 at the University of Virginia. My students have included first and second year graduate students preparing for thesis work in analysis, algebra, or topology, graduate students in various departments in the School of Engineering and Applied Science, and several undergraduate mathematics or physics majors. After a first draft of the manuscript was completed, it was also used for an independent reading course for several undergraduates preparing for graduate school.

While this book is short, comparatively speaking, it does not accomplish it aims through brevity. Arguments are generally presented in detail, and in fact I have tried to firmly keep in mind the reader who may be learning the material on his or her own without the benefit of a formal course or instructor. Since functional analysis is a huge field, I have had to make many omissions with regard to the topics I present. These choices represent, of course, my own preferences, but also my desire to start with the basics and still travel a path through some significant parts of modern functional analysis.

The prerequisites for this book include undergraduate courses in real analysis, linear algebra, and basic point set topology (say, in metric spaces). A modicum of complex analysis is used in a few examples and exercises, and in the proofs of a few results in Chapter 5; in a pinch it is not an essential prerequisite for a student willing to bypass those parts (or take them on faith). With respect to real analysis a good undergraduate level course is essential. Beyond this some familiarity with measure theory and the Lebesgue integral is desirable, but not essential. Save for the last chapter, most of the use of measure theory and Lebesgue integration occurs in limited ways—primarily in examples. An Appendix provides a summary and expository discussion of all that is needed here. I encourage any prospective reader who may feel shaky with these desirable but not essential prerequisites not to be daunted by them. On the basis of my experiences teaching this material I have found that students with no prior exposure to complex analysis or measure theory and Lebesgue integration can nevertheless have a successful experience with the topics presented here.

I have woven a certain amount of historical commentary into the text; this reflects my belief that some understanding of the historical development of any field in mathematics both deepens and enlivens one's appreciation of the subject. The history of functional analysis is filled with interesting characters, many of whom lived and worked during turbulent times in the twentieth century.

Each chapter concludes with an extensive collection of exercises. The purpose of the exercises is to enable the reader to become comfortable with the ideas in the text; to make them his or her own. While most are therefore closely tied to the material being discussed, an occasional exercise is intended to provide an initial step or steps towards a topic not discussed in the text, or to point the way for further exploration. In any case, all are intended to be eminently doable by a student and when advisable are accompanied by a hint.

I would like to express my great appreciation to several friends and colleagues who provided advice and encouragement during the writing of this book. Sheldon Axler, Tom Goebeler, Christopher Hammond, and Bill Ross read substantial portions of the manuscript and provided many helpful comments, as well as suggestions for exercises. Larry Thomas gave useful feedback on the Appendix. Mark Spencer at Springer provided valuable editorial assistance. Julie Riddleberger helped with the illustrations, and patiently answered many TEX questions. I thank Tom Kriete for his enthusiastic support and encouragement throughout all stages of this work. And finally I thank the students in the Functional Analysis course I taught at the University of Virginia in each of the last several years. It was their enthusiastic response to this course that initially got me thinking about writing a functional analysis text, and helped me refine my ideas of what this text should look like.

Charlottesville, Virginia   *Barbara D. MacCluer*
                                                                              July 2008

Contents

Chapter 1
Hilbert Space Preliminaries

It seems to me not useless to indicate interest in a study of sets composed of functions....
J. Hadamard, International Congress of Mathematicians, Zürich, 1897.

Functional analysis developed in the late nineteenth and early twentieth centuries, during a period in which there was a general interest in abstraction, axiomatization, and unification across all fields of mathematics. This unification meant that objects that behaved according to a common set of rules were viewed as "the same," even if they consisted of rather different elements. A core idea in functional analysis is to treat functions as "points" or "elements" in some sort of abstract space, so that instead of working with individual functions (the tradition in classical analysis), we deal with functions as points in a space endowed with some kind of overall structure. The structure of the space itself is emphasized over properties of individual elements in the space. This viewpoint, accompanied by an axiomatization of the new spaces to be considered, was an integral step in the process of transferring familiar concepts in finite-dimensional Euclidean space to (typically infinite-dimensional) "function spaces."

While important contributions to the beginnings of functional analysis were made by individuals of various nationalities, the most readily identifiable schools of work in the early history of the subject were in France, Italy, and Germany. In France, one of the notable contributors to the initial development of functional analysis was Maurice Fréchet, whose 1906 doctoral dissertation is a landmark paper in the subject. In this work, which was extremely influential in both functional analysis and point set topology, Fréchet began the study of abstract spaces of functions. In particular, he defined the notion of a metric space (which he called "(E)" spaces, from the French "écart" meaning distance), and included a discussion of examples of metric spaces where the points in the space were functions. In Fréchet's work one can clearly see the influence of his advisor Jacques Hadamard. In an address to the International Congress of Mathematicians in 1897, Hadamard proposed a study of what would now be termed set-theoretic topology. A quote from this address introduces this chapter; his student Fréchet took up the challenge put forth there.

In this chapter we describe the basic kinds of spaces which will interest us, with a particular emphasis on Hilbert spaces, which are rich in geometric structure. In simplest terms, the idea behind a Hilbert space is to generalize the familiar Euclidean

B.D. MacCluer, *Elementary Functional Analysis*, DOI 10.1007/978-0-387-85529-5_1,
© Springer Science+Business Media, LLC 2009

spaces $\mathbb{R}^n$ or $\mathbb{C}^n$, preserving as much as possible the geometric results in these finite-dimensional settings.

1.1 Normed Linear Spaces

A modern-seeming, axiomatic, definition of vector spaces goes back to the Italian mathematician Giuseppe Peano, in 1888. A vector space is an algebraic object; to introduce such analytic notions as convergence or continuity in a vector space we must provide our vector space with additional structure. This brings us to the concept of a normed linear space, which is a vector space with a norm.

Definition 1.1. Let X be a vector space over either the scalar field $\mathbb{R}$ of real numbers or the scalar field $\mathbb{C}$ of complex numbers. Suppose we have a function $\|\cdot\| : X \to [0, \infty)$ such that

(1) $\|x\| = 0$ if and only if $x = 0$,
(2) $\|x+y\| \leq \|x\| + \|y\|$ for all $x, y \in X$, and
(3) $\|\alpha x\| = |\alpha|\|x\|$ for all scalars α and vectors x.

We call $(X, \|\cdot\|)$ a *normed linear space*.

Property (2) is called the *triangle inequality*, and property (3) is referred to as *homogeneity*. The *reverse triangle inequality*,

$$\|x+y\| \geq |\|x\| - \|y\||$$

follows easily from (2); see Exercise 1.1.

We give some examples of normed linear spaces. In these examples we won't give the details of the verification that the norm satisfies these defining properties. This verification is straightforward in some cases, while in others it may already be known to the reader or will be outlined in an exercise.

Example 1.2. Let $X = \mathbb{C}^n \equiv \{(z_1, z_2, \ldots, z_n) : z_j \in \mathbb{C}\}$ with

$$\|(z_1, z_2, \ldots, z_n)\| = \left(\sum_{j=1}^{n} |z_j|^2\right)^{\frac{1}{2}};$$

this is called the *Euclidean norm*. The Euclidean space $\mathbb{R}^n$ is similarly defined; in this case we restrict to real scalars.

Example 1.3. Let $X = \mathbb{C}^n$ with $\|(z_1, z_2, \ldots, z_n)\| = \max\{|z_j| : 1 \leq j \leq n\}$.

Example 1.4. Let $Y = [0,1]$, or more generally any compact Hausdorff space, and let $C(Y)$ be the vector space of continuous, complex-valued functions on Y, under pointwise addition and scalar multiplication. Define a norm on $C(Y)$ by $\|f\| = \max\{|f(y)| : y \in Y\}$. This (specifically $C[a,b]$, endowed with the metric which defines the distance between functions f and g to be $\max_{a \leq x \leq b} |f(x) - g(x)|$), was one of the important examples that Fréchet put forth in his 1906 dissertation.

Example 1.5. Choose a value of $p \geq 1$, and let $\ell^p = \ell^p(\mathbb{N})$ denote the set of all sequences $\{a_n\}_{n=1}^{\infty}$ of complex numbers (indexed by the positive integers $\mathbb{N}$) for which $\sum_1^{\infty} |a_n|^p < \infty$. In our notation for a sequence we will often abbreviate $\{a_n\}_{n=1}^{\infty}$ by $\{a_n\}_1^{\infty}$ or even just $\{a_n\}$. Define the norm of $\{a_n\} \in \ell^p$ by

$$\|\{a_n\}\|_p \equiv \left(\sum_1^{\infty} |a_n|^p \right)^{1/p}.$$

We can include the choice $p = \infty$ by modifying this definition in the expected way:

$$\ell^{\infty} = \{\{a_n\}_1^{\infty} : \sup_n |a_n| < \infty\}$$

and

$$\|\{a_n\}\|_{\infty} = \sup_n |a_n|.$$

For $p = 1$ and $p = \infty$ the triangle inequality is easily verified; for $1 < p < \infty$ it goes by the name of Minkowski's inequality, in honor of Hermann Minkowski who first studied the analogue of this ℓ^p-norm on the space $\mathbb{R}^n$.

Example 1.6. We can generalize the last example as follows. Consider a positive measure space $(Y, \mathfrak{M}, \mu)$, where Y is a set, $\mathfrak{M}$ is a σ-algebra of subsets of Y, and μ is a positive measure. Choose $1 \leq p < \infty$, and denote by $L^p(Y, \mu)$ the collection of all equivalence classes of $\mathfrak{M}$-measurable functions on Y with

$$\int_Y |f|^p d\mu < \infty,$$

normed by

$$\|f\|_p = \left(\int_Y |f|^p d\mu \right)^{\frac{1}{p}}$$

(the integral in this definition is the Lebesgue integral). Minkowski's inequality (for integrals) provides the proof that the norm satisfies the triangle inequality. We also define $L^{\infty}(X, \mu)$ to be all equivalence classes of essentially bounded measurable functions, normed by $\|f\|_{\infty} = \text{ess sup} |f|$, the essential supremum of f. Of particular interest to us will be the space $L^p[0, 1] = L^p([0, 1], dx)$ with respect to Lebesgue measure dx on the real line.

For the reader unfamiliar with the concepts in the preceding example, the Appendix provides a summary of the relevant definitions and results from real analysis. The use of the Lebesgue integral in the definition of the L^p spaces is important, and in writing a history of functional analysis, Jean Dieudonné [10] states

> ...it is likely that progress in Functional Analysis might have been appreciably slowed down if the invention of the Lebesgue integral had not appeared, by a happy coincidence, exactly at the beginning of Hilbert's work...(pp. 119–120).

replacing, what Dieudonné calls "the horrible and useless so-called Riemann integral." In Example 1.6, the particular choice $Y = \mathbb{N}$ and $\mu = $ counting measure on the subsets of $\mathbb{N}$ gives the space ℓ^p of Example 1.5; see Sections A.2 and A.3 in the Appendix for more details.

Example 1.7. Fix a sequence $\{\beta(n)\}_{n=0}^{\infty}$ of positive numbers with $\beta(0) = 1$ and

$$\lim_{n \to \infty} \beta(n)^{1/n} \geq 1. \tag{1.1}$$

The reason for this last restriction will be made clear shortly, but for right now notice that defining $\beta(n) = (n+1)^a$ for some fixed real number a will give an allowable choice. Define the weighted sequence space ℓ_β^2 to consist of all sequences $\{a_n\}_0^\infty$ with

$$\sum_{n=0}^{\infty} |a_n|^2 \beta(n)^2 < \infty,$$

where the norm of $\{a_n\}_0^\infty$ is defined to be

$$\left(\sum_{n=0}^{\infty} |a_n|^2 \beta(n)^2 \right)^{1/2}.$$

From one perspective these weighted sequence spaces can be thought of simply as $L^2(X, \mathfrak{M}, \mu)$ for $X = \mathbb{N}_0 \equiv \{0\} \cup \mathbb{N}$, $\mathfrak{M}$ the collection of all subsets of $\mathbb{N}_0$, and μ the measure that assigns to each point n of $\mathbb{N}_0$ the mass $\beta(n)^2$, so that we have a special case of the example discussed in Example 1.6. In particular, the general version of Minkowski's inequality gives the triangle inequality in ℓ_β^2. (See Exercise 1.6 for a more elementary approach.)

The requirement in Equation (1.1) allows us to offer a second perspective on the spaces ℓ_β^2, and the interplay between the two perspectives endows these examples with a particular richness. Associate to a sequence $\{a_n\}_0^\infty$ in ℓ_β^2 the power series $\sum_{n=0}^{\infty} a_n z^n$. The radius of convergence of this series is at least one (see Exercise 1.9), and thus the series converges to an analytic function on the unit disk $\mathbb{D} = \{z \in \mathbb{C} : |z| < 1\}$. This suggests that we may want to identify ℓ_β^2, a space of sequences, with the vector space

$$\{f = \sum_0^{\infty} a_n z^n \text{ analytic in } \mathbb{D} : \sum_0^{\infty} |a_n|^2 \beta(n)^2 < \infty\}.$$

In the latter guise, the space is referred to as a *weighted Hardy space* and denoted $H^2(\beta)$; the case $\beta(n) = 1$ for all n gives the *Hardy space* H^2. In the next chapter we will have the language needed to make precise the properties of this identification, but for the moment we simply observe that the map sending $\{a_n\}_0^\infty$ to $f = \sum_0^\infty a_n z^n$ is one-to-one (by uniqueness of power series) and onto $H^2(\beta)$ by definition, and we will regard $H^2(\beta)$ as normed so that this mapping preserves norms.

Example 1.8. Let Ω be a nonempty open set in $\mathbb{C}$. Denote the collection of all bounded analytic functions on Ω by $H^\infty(\Omega)$, and introduce a norm on $H^\infty(\Omega)$ by $\|f\| = \sup\{|f(z)| : z \in \Omega\}$.

The norms in Examples 1.3, 1.4, 1.8, and the ℓ^∞ norm in Example 1.5, are all referred to as the "supremum norm," and when needed for clarity will be written as $\|\cdot\|_\infty$. The L^∞ norm in Example 1.6 is called the essential supremum norm; it is also written $\|\cdot\|_\infty$.

Definition 1.9. A *metric space* is a set X with a function $d(\cdot,\cdot) : X \times X \to [0,\infty)$ satisfying, for x,y, and z in X,

(1) $d(x,y) = 0$ if and only if $x = y$,
(2) $d(x,y) = d(y,x)$, and
(3) $d(x,y) + d(y,z) \geq d(x,z)$.

The third property is referred to as the triangle inequality.

On any metric space (X,d) there is an associated topology. The open balls are the sets of the form $B(a,r) \equiv \{x : d(x,a) < r\}$, where $r > 0$. Every open set is the union of some collection of open balls. It is easy to see that if X is a normed linear space, we may define a metric on X by defining $d(x,y) = \|x-y\|$. With the metric topology in place on X, continuity of certain basic mappings can be addressed. For example, it is easy to check that the function $\|\cdot\| : X \to [0,\infty)$ is continuous; see Exercise 1.8 for this and other elementary results.

About eight years after Fréchet's seminal work in 1906, Felix Hausdorff wrote a text that presented a thoroughly modern definition of metric space and defined the fundamental idea of a Cauchy sequence, which we recall next.

Definition 1.10. Let X be a metric space. A sequence $\{x_n\}$ in X is said to be a *Cauchy sequence* if it has the following property: Given any $\varepsilon > 0$ there exists N such that if $n, m \geq N$, then $d(x_n, x_m) < \varepsilon$.

Definition 1.11. A metric space is said to be *complete* if every Cauchy sequence in X converges in X.

Definition 1.12. Let X be a normed linear space. If X is complete in the metric d defined from the norm by $d(x,y) = \|x-y\|$, we call X a *Banach space*.

All of the above examples of normed linear spaces are Banach spaces. We will not stop to prove this now, but we do make a couple of observations. The statement that the space $L^p(Y,\mu)$ is complete for any $1 \leq p \leq \infty$ and any positive measure space (Y,μ) goes by the name of the *Riesz–Fischer theorem*. In its full generality it is a deep result of real analysis (see also the discussion in Section 1.5 below and in Section A.3 of the Appendix). Notice that this general class of examples includes the ℓ^p spaces and weighted sequences spaces as special cases (see Exercise 1.6 for a more elementary approach), as well as the finite-dimensional spaces in Examples 1.2 and 1.3. In Exercise 1.2 the reader is asked to provide a proof of completeness for the spaces in Example 1.4, and a similar argument can be used for the space $H^\infty(\Omega)$ of Example 1.8. You can get an example of a normed linear space which is not a Banach space by taking a nonclosed subspace of a Banach space; see for example Exercise 1.3. (A *subspace* of a vector space V is a subset of V which is itself a vector space under the same addition and scalar multiplication operations.)

Banach spaces are named in honor of the Polish mathematician Stefan Banach, a dominating figure in the birth of functional analysis, who wrote a fundamentally important book called *Opérations Linéaires* in 1932. In this book (which had its beginnings in Banach's 1920 doctoral thesis) many of the properties of complete normed linear spaces are developed. Banach calls these spaces "spaces of type (B)," perhaps in the hope they would eventually be known as "Banach spaces"[1] This is precisely what happened, with the terminology "Banach space" making its formal appearance in Fréchet's text *Les Espaces Abstraits* [13].

Hugo Steinhaus, Banach's teacher and collaborator, writes in a 1963 memoir of Banach that Banach's axiomatic definition of a complete normed linear space provided precisely the right level of generality; broad enough to encompass a wide variety of natural examples, but not so general as to permit only uninteresting theorems:

> His foreign competitors in the theory of linear operations either dealt with spaces that were too general, and that is why they either obtained only trivial results, or assumed too much about those spaces, which restricted the extent of the applications to a few and artificial examples — Banach's genius reveals itself in finding the golden mean. This ability of hitting the mark proves that Banach was born a high class mathematician ([44], p. 12).

In fact, a few months after Banach set down the axioms for a normed linear space, the American Norbert Wiener independently gave nearly the same definition, and for a short while the terminology "Banach–Wiener spaces" was used. However, as Wiener's interest in the area did not continue, these spaces, in Wiener's words, became "quite justly named after Banach alone ([46], p. 60)."

Hilbert spaces, which we turn to now, are Banach spaces with some additional structure, coming from the presence of an inner product.

Definition 1.13. Let X be a vector space over $\mathbb{C}$. An *inner product* is a map $\langle \cdot, \cdot \rangle : X \times X \to \mathbb{C}$ satisfying, for x, y, and z in X and scalars $\alpha \in \mathbb{C}$,

(1) $\langle x, y \rangle = \overline{\langle y, x \rangle}$ for all x, y in X,

(2) $\langle x, x \rangle \geq 0$, with $\langle x, x \rangle = 0$ (if and) only if $x = 0$,

(3) $\langle x + y, z \rangle = \langle x, z \rangle + \langle y, z \rangle$, and

(4) $\langle \alpha x, y \rangle = \alpha \langle x, y \rangle$.

Some comments on this definition are in order. The bar in (1) denotes complex conjugation. Property (2) is referred to as "positive-definiteness," and the adjective "Hermitian" is used for property (1). The parenthetical "if" statement in (2) need not be included in the definition, as it follows from the other parts since $\langle 0, 0 \rangle = \langle 2 \cdot 0, 0 \rangle = 2 \langle 0, 0 \rangle$. An inner product is linear in the first slot and conjugate linear in

---

[1] Though this interpretation of Banach's choice of notation is widely repeated, V.D. Milman, in writing about Banach, says, "In his book...Banach denotes operators by the letter A. These were the initial objects of study, and the complete normed spaces on which they operated were denoted by the Latin letter B. That was natural, and there is no indication that he was 'hinting' at his own name by using that letter" ([32], p. 228).

the second ($\langle x, \alpha y + z \rangle = \overline{\alpha} \langle x, y \rangle + \langle x, z \rangle$), so the defining properties are encapsulated by saying that an inner product is a Hermitian, positive definite, sesquilinear form (sesquilinear from the Latin for "$1\frac{1}{2}$" linear). The reader is cautioned that some authors (in physics, for example) define the inner product to be linear in the second slot, and conjugate linear in the first.

A standard example is to define an inner product on $L^2(X, \mu)$ for a positive measure space (X, μ) by

$$\langle f, g \rangle = \int_X f\overline{g} \, d\mu.$$

This general framework includes, as special cases, the example $\mathbb{C}^n$ with

$$\langle (z_1, z_2, \dots, z_n), (w_1, w_2, \dots, w_n) \rangle = \sum_{j=1}^{n} z_j \overline{w_j},$$

the example ℓ^2 of all square summable sequences with

$$\langle (z_1, z_2, \dots), (w_1, w_2, \dots) \rangle = \sum_{j=1}^{\infty} z_j \overline{w_j},$$

and the weighted analogues ℓ_β^2 with

$$\langle (z_0, z_1, \dots), (w_0, w_1, \dots) \rangle = \sum_{j=0}^{\infty} z_j \overline{w_j} \beta(j)^2.$$

The first two are obtained by taking X to be, respectively, $\{1, 2, \dots, n\}$ or $\mathbb{N}$, with μ equal to counting measure. In the case of weighted sequence spaces, $X = \mathbb{N}_0$ and μ assigns mass $\beta(n)^2$ to the set $\{n\}$.

Any inner product satisfies an important inequality, called the Cauchy–Schwarz inequality, which we describe next.

Proposition 1.14. *If $\langle \cdot, \cdot \rangle$ is an inner product on a vector space X, then for all x and y in X we have*

$$|\langle x, y \rangle|^2 \le \langle x, x \rangle \langle y, y \rangle.$$

In this general form, the Cauchy–Schwarz inequality is due to John von Neumann (1930), who is often credited with the "axiomatization" of Hilbert spaces (defined below). Earlier versions of Proposition 1.14, for specific settings, go back to Cauchy, Bunyakowsky, and Schwarz, and the Cauchy–Schwarz inequality is sometimes referred to as the Cauchy–Bunyakowsky–Schwarz inequality.

One particularly simple proof of Proposition 1.14 is outlined in Exercise 1.7. As an important application of Proposition 1.14, we show next how any inner product defines a norm.

Proposition 1.15. *If $\langle \cdot, \cdot \rangle$ is an inner product on a vector space X, then*

$$\|x\| \equiv \langle x, x \rangle^{\frac{1}{2}}$$

is a norm on X.

Proof. We will check the triangle inequality, and leave the verification of the other norm properties to the reader. Using the linearity of the inner product we have

$$\begin{aligned}
\|x+y\|^2 &= \langle x+y, x+y \rangle = \langle x, x \rangle + \langle y, x \rangle + \langle x, y \rangle + \langle y, y \rangle \\
&= \|x\|^2 + 2\mathrm{Re}\langle x, y \rangle + \|y\|^2 \\
&\leq \|x\|^2 + 2|\langle x, y \rangle| + \|y\|^2 \\
&\leq \|x\|^2 + 2\|x\|\|y\| + \|y\|^2 \\
&= (\|x\| + \|y\|)^2
\end{aligned}$$

where Re z denotes the real part of a complex number z, and we have used the Cauchy–Schwarz inequality in the penultimate step.   □

Definition 1.16. A (complex) *Hilbert space* $\mathscr{H}$ is a vector space over $\mathbb{C}$ with an inner product such that $\mathscr{H}$ is complete in the metric

$$d(x,y) = \|x-y\| = \langle x-y, x-y \rangle^{\frac{1}{2}}.$$

Any space $L^2(X,\mu)$ as described above is thus an example of a Hilbert space, since we have already observed that $L^2(X,\mu)$ is a Banach space under the norm $\|f\|_2 = (\int_X |f|^2 d\mu)^{\frac{1}{2}}$ which we recognize as $\langle f, f \rangle^{\frac{1}{2}}$.

There are various anecdotes, of dubious validity, about David Hilbert and the terminology "Hilbert space." Steve Krantz, writing in *Mathematical Apocrypha* [27] says

> It is said that, late in his life, Hilbert was reading a paper and got stuck at one point. He went to his colleague in the office next door and queried, "What is a Hilbert space?" (p. 89)

Another version is given by Laurence Young [47]:

> When Weyl presented a proof of the Riesz–Fischer theorem in a Göttingen colloquium, Hilbert went up to the speaker afterward to say, "Weyl, you must just tell me one thing, whatever is a Hilbert space?" (p. 312)

Next we will look at an important example of a Hilbert space where the vectors are certain analytic functions on the unit disk $\mathbb{D} = \{z \in \mathbb{C} : |z| < 1\}$. This example, which uses a few basic results from complex analysis, will prove to be particularly illuminating of several of the fundamental Hilbert space notions.

Example 1.17. The Bergman space $L_a^2(\mathbb{D})$ is the vector space, under pointwise operations, of all *analytic* functions f on $\mathbb{D}$ for which

$$\int_{\mathbb{D}} |f(z)|^2 \frac{dA}{\pi} < \infty,$$

where dA denotes two-dimensional Lebesgue measure (so that dA/π is "normalized area measure" on the unit disk). Of course, every function in $L_a^2(\mathbb{D})$ is (a representative of) an element of the Hilbert space $L^2(\mathbb{D}, dA/\pi)$; we give $L_a^2(\mathbb{D})$ the inner product it inherits from $L^2(\mathbb{D})$:

$$\langle f,g \rangle = \int_{\mathbb{D}} f(z)\overline{g(z)}\frac{dA}{\pi}.$$

Our first goal is to check that the Bergman space is a Hilbert space. How much work must we do? Since we already know that $L^2(\mathbb{D},dA/\pi)$ is a Hilbert space, it will suffice to verify that $L_a^2(\mathbb{D})$ is a *closed* subspace of $L^2(\mathbb{D},dA/\pi)$. Of course, "closed" here refers to the topology on $L^2(\mathbb{D},dA/\pi)$; this is the metric topology induced by the norm. To this end, we need an area mean-value property for analytic functions.

Proposition 1.18. *If f is a analytic function in some closed disk $\overline{B(a,R)}$, then*

$$f(a) = \frac{1}{\pi R^2} \int_{B(a,R)} f\, dA.$$

Proof. As a consequence of Cauchy's integral formula we have the mean value property

$$f(a) = \frac{1}{2\pi} \int_0^{2\pi} f(a+re^{i\theta})d\theta$$

for all $0 < r < R$. Multiplying by r and integrating with respect to r we have

$$\int_0^R rf(a)dr = \int_0^R \int_0^{2\pi} f(a+re^{i\theta})r\frac{d\theta}{2\pi}dr$$

or equivalently

$$f(a)\frac{R^2}{2} = \frac{1}{2\pi}\int_{B(a,R)} fdA$$

as desired. □

From this we get a corollary that gives an upper bound on the value of a function in the Bergman space at a point $w \in \mathbb{D}$ in terms of the norm of f and the distance from w to $\partial\mathbb{D}$, the unit circle.

Corollary 1.19. *Fix $w \in \mathbb{D}$. For every $f \in L_a^2(\mathbb{D})$ we have*

$$|f(w)| \le \frac{1}{1-|w|}\|f\|_{L_a^2(\mathbb{D})}.$$

Proof. Let $0 < r < 1 - |w|$ so that the closed disk $\overline{B(w,r)}$ is contained in $\mathbb{D}$. Using Proposition 1.18 and Hölder's inequality we have

$$|f(w)| = \left| \frac{1}{\pi r^2} \int_{B(w,r)} f\, dA \right|$$

$$\le \frac{1}{\pi r^2} \int_{B(w,r)} |f|\, dA$$

$$\le \frac{1}{\pi r^2} \left(\int_{B(w,r)} 1\, dA \right)^{1/2} \left(\int_{B(w,r)} |f|^2\, dA \right)^{1/2}$$

$$\leq \frac{1}{\pi r^2} \sqrt{\pi r^2} \sqrt{\pi} \left(\int_{\mathbb{D}} |f|^2 \, \frac{dA}{\pi} \right)^{1/2}$$

$$= \frac{1}{r} \|f\|_{L^2_a(\mathbb{D})}.$$

This calculation holds for any $r < 1 - |w|$, so letting r increase to $1 - |w|$ yields the desired conclusion. $\square$

To show that the Bergman space is a Hilbert space, we will use, in addition to Corollary 1.19, a result from real analysis that says if a sequence $\{f_n\}$ in $L^2(\mathbb{D}, dA/\pi)$ converges in the L^2-norm to a limit f, then some subsequence $\{f_{n_k}\}$ converges pointwise almost everywhere (dA/π) to f; see, for example, p. 74 in [40].

Theorem 1.20. *The Bergman space* $L^2_a(\mathbb{D})$ *is a Hilbert space.*

Proof. As we have discussed, we need only show that $L^2_a(\mathbb{D})$ is a closed subspace of $L^2(\mathbb{D}, dA/\pi)$. That $L^2_a(\mathbb{D})$ is a subspace is immediate. To see that it is closed, suppose we have a sequence $\{f_n\}$ of functions in $L^2_a(\mathbb{D})$ with $f_n \to f$ in $L^2(\mathbb{D}, dA/\pi)$. Our task is to show that f must be in $L^2_a(\mathbb{D})$; that is, that the limit function f is analytic (or more precisely, has an analytic representative). On the one hand, from the remark preceding the theorem, we know that convergence of f_n to f in the norm of $L^2(\mathbb{D}, dA/\pi)$ implies some subsequence $\{f_{n_k}\}$ converges pointwise almost everywhere (dA/π) to f. On the other hand, by Corollary 1.19 we have, for any closed disk $\overline{B(0,r)} \subset \mathbb{D}$ and all z in this closed disk,

$$|f_n(z) - f_m(z)| \leq \frac{1}{1-r} \|f_n - f_m\|_{L^2_a(\mathbb{D})}.$$

This says that the sequence $\{f_n\}$ is uniformly Cauchy on $\overline{B(0,r)}$ and, by Morera's theorem from complex analysis, f_n converges uniformly on $B(0,r)$ to an analytic function g on $B(0,r)$. This holds for all $r < 1$ and thus our pointwise limit f must agree almost everywhere with an analytic function, i.e., we may choose f to be analytic in $L^2(\mathbb{D}, dA/\pi)$. This is precisely the desired conclusion that f is in the Bergman space $L^2_a(\mathbb{D})$. $\square$

There are L^p, $p \neq 2$, versions of the Bergman space; see Exercise 1.11 for the definition and some basic properties.

1.2 Orthogonality

A Banach space is a complete normed linear space and a Hilbert space is a complete inner product space. The presence of an inner product permits the all-important geometric notion of orthogonality, which says in turn that Hilbert spaces behave in many ways as generalizations of finite-dimensional Euclidean space, where one can talk about angles and projections, for example.

Definition 1.21. Given vectors f, g in a Hilbert space $\mathcal{H}$, we say that f is *orthogonal to* g, written $f \perp g$, if $\langle f, g \rangle = 0$. For sets A and B in $\mathcal{H}$ we write $A \perp B$ if $\langle f, g \rangle = 0$ for all $f \in A$ and $g \in B$. Finally, $A^\perp$ is the set of all vectors $f \in \mathcal{H}$ such that $f \perp g$ for all g in A; for any set A this is always a subspace of $\mathcal{H}$, moreover since $A^\perp = \cap_{a \in A} \{a\}^\perp$, $A^\perp$ is a closed subspace by continuity of the inner product (see Exercise 1.8).

It should be clear that $A \cap A^\perp = \{0\}$. (Why?)

Some authors use the terminology "linear manifold" for a linear subspace that is not necessarily closed, and reserve the term "subspace" for a closed linear manifold. We will not do so, but instead use the adjective "closed" when it applies. An example of a subspace which is not closed is the set of all sequences in ℓ^2 with finitely many nonzero terms.

The next result, aptly called the *Pythagorean theorem*, is easily verified by writing the norm in terms of the inner product and expanding. The details are left to the reader.

Proposition 1.22. *If* $f_1, f_2, \ldots, f_n$ *are pairwise orthogonal vectors in a Hilbert space, then*

$$\|f_1 + f_2 + \cdots + f_n\|^2 = \|f_1\|^2 + \|f_2\|^2 + \cdots + \|f_n\|^2.$$

In general, for any vectors f and g in a Hilbert space we have

$$\|f + g\|^2 = \|f\|^2 + 2\operatorname{Re} \langle f, g \rangle + \|g\|^2$$

and

$$\|f - g\|^2 = \|f\|^2 - 2\operatorname{Re} \langle f, g \rangle + \|g\|^2.$$

The *parallelogram equality* is then obtained:

$$\|f + g\|^2 + \|f - g\|^2 = 2\|f\|^2 + 2\|g\|^2.$$

Its name comes from picturing the relationship for vectors in, say, $\mathbb{R}^2$; see Figure 1.1.

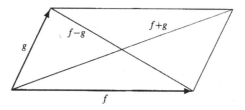

FIGURE 1.1: The parallelogram equality

In any inner product space, the inner product can be recovered from the norm:

$$\langle f, g \rangle = \frac{1}{4} \left(\|f+g\|^2 - \|f-g\|^2 + i\|f+ig\|^2 - i\|f-ig\|^2 \right). \tag{1.2}$$

This is called the *polarization identity*, and it is verified by a straightforward calculation. It can be written, and perhaps more easily remembered, as

$$\langle f, g \rangle = \frac{1}{4} \sum_{k=0}^{3} i^k \|f + i^k g\|^2.$$

Perhaps surprisingly, given a normed linear space in which the parallelogram equality holds, there is an inner product that gives the norm. See Exercise 1.14 for an outline of how to show this result, which is due to P. Jordan and J. von Neumann. This exercise gives the best known of many (hundreds!) of ways of characterizing those normed linear spaces that are in fact inner product spaces. For much more on this subject, the reader is referred to [1].

1.3 Hilbert Space Geometry

A *convex set* in a vector space V is a subset S of V with the property that whenever a, b are in S, so is $ta + (1-t)b$ for any $0 \le t \le 1$. Clearly every subspace is convex, every ball in a normed linear space is convex, and any translate $x + S \equiv \{x + s : s \in S\}$ of a convex set S is convex. The next result, which we will refer to as the *nearest point property*, is a key step in obtaining our main theorem on Hilbert space geometry.

Proposition 1.23 (Nearest Point Property). *Every nonempty, closed convex set K in a Hilbert space $\mathscr{H}$ contains a unique element of smallest norm. Moreover, given any $h \in \mathscr{H}$, there is a unique k_0 in K such that*

$$\|h - k_0\| = \operatorname{dist}(h, K) \equiv \inf\{\|h - k\| : k \in K\}.$$

Proof. We begin with a proof of the first statement. The parallelogram equality says that for any vectors x, y in $\mathscr{H}$,

$$\left\| \frac{x-y}{2} \right\|^2 = \frac{1}{2}(\|x\|^2 + \|y\|^2) - \left\| \frac{x+y}{2} \right\|^2.$$

If $d \equiv \inf\{\|y\| : y \in K\}$, then we may find a sequence of vectors $\{x_n\}$ in K with $\|x_n\| \to d$. Thus for any n, m we have

$$\left\| \frac{x_n - x_m}{2} \right\|^2 = \frac{1}{2}(\|x_n\|^2 + \|x_m\|^2) - \left\| \frac{x_n + x_m}{2} \right\|^2,$$

where, by convexity, $\frac{1}{2}x_n + \frac{1}{2}x_m$ is in K, so that

$$\left\|\frac{x_n + x_m}{2}\right\|^2 \geq d^2.$$

Thus we have

$$0 \leq \|x_n - x_m\|^2 \leq 2(\|x_n\|^2 + \|x_m\|^2) - 4d^2.$$

This tells us that $\{x_n\}$ is a Cauchy sequence, and by completeness it must converge to some $x \in \mathcal{H}$. Since K is closed, $x \in K$. Continuity of the norm says that $\|x_n\| \to \|x\|$, so $\|x\| = d$. This gives us the existence part of the first statement.

For uniqueness, suppose $\|z\| = \|x\| = d$ for some z in K. Consider $\frac{1}{2}x + \frac{1}{2}z \in K$. Since we must have

$$\left\|\frac{x+z}{2}\right\| \geq d$$

the parallelogram equality again says

$$\left\|\frac{x-z}{2}\right\|^2 = \frac{1}{2}(\|x\|^2 + \|z\|^2) - \left\|\frac{x+z}{2}\right\|^2 = d^2 - \left\|\frac{x+z}{2}\right\|^2 \leq 0,$$

which forces $x = z$. This completes the proof of the first statement.

The second statement is obtained by translation. To find the unique point in K closest to a given h in $\mathcal{H}$, first find the unique point x in the convex set $K - h$ of minimal norm. Its translate $x + h$ is the desired point. $\qquad\square$

The arguments used in this proof are basically those of the Hungarian mathematician Frederic Riesz, another important contributor during the early period of functional analysis. We will continue to see his name attached to quite a few of the results discussed in this book.

The nearest point property is quite rigid—it fails to be true if we omit either the requirement that K be closed or convex, or change "Hilbert space" to "Banach space" in the statement. The interested reader can provide examples to illustrate this.

We will get a lot of mileage out of the next result, called the *projection theorem*, whose proof uses the nearest point property. Our presentation follows that of [40].

Theorem 1.24 (Projection Theorem). *Let M be a closed subspace of a Hilbert space $\mathcal{H}$. There is a unique pair of mappings $P : \mathcal{H} \to M$ and $Q : \mathcal{H} \to M^\perp$ such that $x = Px + Qx$ for all $x \in \mathcal{H}$. Furthermore, P and Q have the following additional properties:*

(a) $x \in M \Longrightarrow Px = x$ and $Qx = 0.$
(b) $x \in M^\perp \Longrightarrow Px = 0$ and $Qx = x.$
(c) *Px is the closest vector in M to x.*
(d) *Qx is the closest vector in $M^\perp$ to x.*
(e) $\|Px\|^2 + \|Qx\|^2 = \|x\|^2$ *for all x.*
(f) *P and Q are linear maps.*

Proof. First we define P as follows: For every $x \in \mathcal{H}$, let Px be the unique closest point to x in the (closed convex) set M; here we are using the nearest point property of Proposition 1.23. Uniqueness says that P is well-defined. Moreover, $P : \mathscr{H} \to M$, and if $x \in M$, $Px = x$. Define $Qx = x - Px$ so that Q is uniquely defined on $\mathscr{H}$, $Px + Qx = x$ for all x, and if $x \in M$, $Qx = 0$.

We next show that $Qx \in M^{\perp}$ for all x. It suffices to show that $\langle x - Px, m \rangle = 0$ for all $m \in M$. Clearly it is enough to check this for unit vectors in M. Fix $m \in M$, $\|m\| = 1$. Consider $Px + \alpha m \in M$ for α any complex number. Since Px is the closest point to x in M

$$\|x - (Px + \alpha m)\| \geq \|x - Px\|.$$

Writing $z = x - Px$ we have

$$\|z\|^2 + |\alpha|^2 \|m\|^2 - \langle \alpha m, z \rangle - \langle z, \alpha m \rangle \geq \|z\|^2.$$

This is true for all α complex, and choosing $\alpha = \langle z, m \rangle$ we see that $\alpha = 0$. This says

$$0 = \alpha = \langle z, m \rangle = \langle x - Px, m \rangle,$$

which verifies the claim and proves that $Qx \in M^{\perp}$ as desired.

If $x \in M^{\perp}$, then $x - Qx \in M^{\perp}$, since $M^{\perp}$ is a subspace. But also $x - Qx = Px \in M$, and $M \cap M^{\perp} = \{0\}$, so $x \in M^{\perp}$ implies $x = Qx$ and $Px = 0$. Furthermore, for any $x \in \mathscr{H}$,

$$\|x\|^2 = \langle Px + Qx, Px + Qx \rangle = \|Px\|^2 + \|Qx\|^2,$$

since $Qx \in M^{\perp}$ and $Px \in M$.

Next we check the linearity of the maps P and Q. Let x, y be arbitrary vectors in $\mathscr{H}$. We want to show $P(x + y) = Px + Py$ and similarly for Q. Since $x = Px + Qx$, $y = Py + Qy$ and $x + y = P(x + y) + Q(x + y)$ we see that

$$Px + Qx + Py + Qy = P(x + y) + Q(x + y)$$

so that

$$Qx + Qy - Q(x + y) = P(x + y) - Px - Py.$$

The vector on the left side of the last line lies in $M^{\perp}$ and the vector on the right side in M; this forces both to be 0, and we have our desired conclusions. The statements $P(\alpha x) = \alpha Px$ and $Q(\alpha x) = \alpha Qx$ are proved similarly.

There are two remaining parts to the proof. We must show that Qx is the closest vector in $M^{\perp}$ to x, and verify the uniqueness statement for P and Q. The first of these goes as follows: Let $x \in \mathscr{H}$, and suppose $y \in M^{\perp}$. Since $Qx \in M^{\perp}$ for all x,

$$\|x - y\|^2 = \|Px + Qx - y\|^2 = \|Px\|^2 + \|Qx - y\|^2;$$

this is clearly minimized if $y = Qx$. So Qx is the closest vector to x in $M^{\perp}$.

The uniqueness of P and Q with the specified properties is easy: if $P, P' : \mathscr{H} \to M$ and $Q, Q' : \mathscr{H} \to M^{\perp}$ with $Px + Qx = x = P'x + Q'x$ for all x then $Px - P'x = Q'x - Qx$; the common value must be 0 since $M \cap M^{\perp} = \{0\}$.   □

Figure 1.2 illustrates the projections P and Q from the projection theorem.

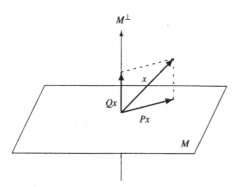

FIGURE 1.2: The projections P and Q

The linear maps P and Q in the projection theorem are called the orthogonal projections of $\mathcal{H}$ onto M and $M^\perp$, respectively. The notation $\mathcal{H} = M \oplus M^\perp$ is commonly used to encapsulate the statement of the projection theorem.

We end this section with a simple, but useful, corollary of the projection theorem, whose proof is left to the reader.

Corollary 1.25. *If M is a closed, proper, subspace of $\mathcal{H}$, then there exists a non-zero vector y in $\mathcal{H}$ with $y \perp M$.*

As a consequence of this corollary, note that one can show that a closed subspace M is all of $\mathcal{H}$ by showing that there is no nonzero vector y in $\mathcal{H}$ with $y \perp M$.

1.4 Linear Functionals

Definition 1.26. If X is a normed linear space over $\mathbb{C}$, a *linear functional* on X is a map $\Lambda : X \to \mathbb{C}$ satisfying $\Lambda(\alpha x + \beta y) = \alpha \Lambda(x) + \beta \Lambda(y)$ for all vectors x and y in X and all scalars α and β.

Hadamard in 1903, and his student Fréchet in 1904–05, began to investigate the *continuous linear functionals* on various function spaces. Hadamard, for example, described the linear functionals on $C[a,b]$ as having the form

$$\Lambda(f) = \lim_{n \to \infty} \int f(x) \Phi_n(x) dx$$

for a sequence of continuous functions Φ_n and, in a letter to Fréchet in 1904, proposed the term "functional" for these "functions of functions." When the function space under investigation was a Hilbert space, work done independently by Fréchet and Riesz gave a particularly pleasant and important characterization of these linear functionals, as we will soon see. We begin with a definition.

Definition 1.27. A *bounded linear functional* on a normed linear space X is a linear functional $\Lambda : X \to \mathbb{C}$ for which there exists a finite constant C satisfying $|\Lambda(x)| \leq C\|x\|$ for all $x \in X$.

Anticipating notation somewhat, we write

$$\|\Lambda\| \equiv \sup\{|\Lambda(x)| : \|x\| \leq 1\}$$

and refer to this as the norm of Λ; it is easy to check that when we give linear structure to the collection of all bounded linear functionals on any normed linear space by defining the vector operators of addition and scalar multiplication pointwise, $\|\cdot\|$ is indeed a norm on this linear space. More will be said about structure on this space later. Exercise 1.16 gives several equivalent formulations of $\|\Lambda\|$, which we will use without comment in what follows. An easy, but fundamental, observation is that bounded linear functionals are precisely those linear functionals which are continuous.

Proposition 1.28. *If X is a normed linear space, and $\Lambda : X \to \mathbb{C}$ is a linear functional, then the following are equivalent:*

(a) Λ is continuous.
(b) Λ is continuous at 0.
(c) Λ is bounded.

Proof. The implication (a)$\Rightarrow$(b) is trivial, so we look first at (b)$\Rightarrow$(c). Since $\Lambda(0) = 0$ (why?), continuity of Λ at 0 means that given $\varepsilon > 0$ we may find $\delta > 0$ such that if $\|x\| \leq \delta$, then $|\Lambda(x)| < \varepsilon$. Choose such a δ to correspond to $\varepsilon = 1$. Given $x \neq 0$, linearity tells us that

$$|\Lambda(x)| = \left|\Lambda\left(\frac{\|x\|}{\delta} \cdot \frac{x\delta}{\|x\|}\right)\right| \leq \frac{\|x\|}{\delta},$$

which gives the boundedness of Λ with $\|\Lambda\| \leq 1/\delta$. The result of Exercise 1.16 is used here. The proof of (c)$\Rightarrow$(a) is left to the reader. $\square$

In discussing continuity, it is helpful to recall that *any* map $f : X \to Y$, where X is a metric space but Y need only be a topological space, is continuous if and only if given any sequence $\{x_n\}$ in X that converges to a point x_0 in X, the sequence $\{f(x_n)\}$ converges to $f(x_0)$. The "if" direction of this statement need not be true if X is not a metric space; this issue will be discussed further in Section 5.5.

As a consequence of a somewhat more general result that we will prove later, the set of all bounded linear functionals on a normed linear space X is itself a Banach space, under pointwise operations and using the norm just defined. In terminology proposed by Nicolas Bourbaki [2] in 1938, this is called the *dual space* of X and

---

[2] Nicolas Bourbaki is the pseudonym of a "secret" society of mathematicians, nearly all French, formed in 1935. It included among its founding members A. Weil, J. Dieudonné, and H. Cartan. New members were added over time, and one of its rules was that members were to retire at age

is denoted X^*. Our immediate goal is to understand the dual space of a Hilbert space. The next result is called the *Riesz representation theorem*; it was discovered independently by Riesz and Fréchet in 1907. As motivation for the statement, observe that if we fix a vector h_0 in a Hilbert space $\mathcal{H}$, then the map $\Lambda(h) \equiv \langle h, h_0 \rangle$ is clearly linear on $\mathcal{H}$. The Cauchy–Schwarz inequality shows that Λ is bounded with $\|\Lambda\| \leq \|h_0\|$. In fact we have equality, as is easily seen by computing $\Lambda(h_0/\|h_0\|)$ if $h_0 \neq 0$. The Riesz representation theorem provides a converse to these observations.

Theorem 1.29. *Every bounded linear functional Λ on a Hilbert space $\mathcal{H}$ is given by inner product with a (unique) fixed vector h_0 in $\mathcal{H}$: $\Lambda(h) = \langle h, h_0 \rangle$. Moreover, the norm of the linear functional Λ is $\|h_0\|$.*

Proof. Suppose Λ is a bounded linear functional on $\mathcal{H}$. If Λ is identically 0, choose $h_0 = 0$. Otherwise, set

$$M = \ker \Lambda \equiv \{h \in \mathcal{H} : \Lambda(h) = 0\}.$$

Since Λ is linear, M is a subspace of $\mathcal{H}$, and since Λ is continuous, $M = \Lambda^{-1}(0)$ is closed. Note that $M \neq \mathcal{H}$ since we are assuming $\Lambda \neq 0$. Pick a nonzero vector $z \in M^\perp$. By scaling if necessary we may assume $\Lambda(z) = 1$. Consider, for arbitrary $h \in \mathcal{H}$, the vector $\Lambda(h)z - h$ and observe that if we apply Λ to this vector we get 0, i.e., it lies in M. Since z was chosen to lie in $M^\perp$, this says

$$\Lambda(h)z - h \perp z$$

so that for every $h \in \mathcal{H}$,

$$\langle \Lambda(h)z - h, z \rangle = 0.$$

Rearranging this last line we see that $\Lambda(h) = \langle h, z/\|z\|^2 \rangle$, which gives the existence statement with $h_0 = z/\|z\|^2$. Uniqueness is immediate, and since we have already observed that $\|\Lambda\| = \|h_0\|$, we are done. $\square$

What does the proof of this result tell you about the relationship between any two vectors in $(\ker \Lambda)^\perp$ when Λ is a bounded linear functional on $\mathcal{H}$?

Theorem 1.29 says that a Hilbert space is self-dual, i.e., that $\mathcal{H}^* = \mathcal{H}$ in the sense that the map sending h_0 in $\mathcal{H}$ to the bounded linear functional $\langle \cdot, h_0 \rangle$ is an isometry of $\mathcal{H}$ onto its dual space ("isometry" referring to the fact that the norm of the linear functional induced by h_0 is $\|h_0\|$). Notice that we're not asserting linearity for the identification of h_0 with the linear function $\langle \cdot, h_0 \rangle$; why not?

In their 1907 works, Riesz and Fréchet dealt specifically with the Hilbert space $L^2[a,b]$. Shortly thereafter, Riesz considered the natural generalization of his work when he investigated the possibility of describing all bounded linear functionals on $L^p[a,b]$ for $1 \leq p < \infty$, launching the study of L^p spaces as normed linear spaces. In 1909, Riesz identified the set of all bounded linear functionals on $L^p[a,b], 1 \leq p < \infty$

---

50. The society's original purpose was to create an analysis text, but this quickly expanded into a project of much bigger scope. A multivolume *Éléments de mathématique*, now totaling more than 7000 pages and treating many core topics in modern mathematics, has been produced.

with $L^q[a,b]$, where $1/p + 1/q = 1$ (when $p = 1$ we set $q = \infty$). The analogous statement for ℓ^p came a few years earlier, in work of E. Landau; the reader is asked to provide a proof in this case in Exercise 1.17. If we leave the realm of Banach spaces, however, a discussion of bounded linear functionals may become moot. For example, M.M. Day showed in 1940 that there are no continuous linear functionals on $L^p[0,1]$ for $0 < p < 1$ except the trivial functional (which is identically zero). The spaces $L^p[0,1]$ for $0 < p < 1$ are discussed in Exercise 1.30; they are not Banach spaces.

Let us return to our example of the Bergman space $L_a^2(\mathbb{D})$. Observe that Corollary 1.19 says that evaluation at any point $w \in \mathbb{D}$ is a bounded linear functional on the Hilbert space $L_a^2(\mathbb{D})$. By Theorem 1.29, evaluation at w must thus be given by inner product with some fixed vector in $L_a^2(\mathbb{D})$, that is, for each $w \in \mathbb{D}$ there is a function in $L_a^2(\mathbb{D})$, which we will denote $K_w(z)$, satisfying $f(w) = \langle f, K_w \rangle$ for all $f \in L_a^2(\mathbb{D})$. Can we identify K_w? This has a nice answer, which is outlined in Exercise 1.25.

Next we'll interpret the projection theorem when $\mathscr{H} = L^2(\mathbb{D}, dA/\pi)$ and $M = L_a^2(\mathbb{D})$, the Bergman space. Can we find an explicit formula for the orthogonal projection $P : L^2(\mathbb{D}, dA/\pi) \to L_a^2(\mathbb{D})$? A simple lemma will be useful here.

Lemma 1.30. *Let* $P : \mathscr{H} \to M$ *be the orthogonal projection of a Hilbert space* $\mathscr{H}$ *onto a closed subspace* M *of* $\mathscr{H}$. *We have* $\langle f, Pg \rangle = \langle Pf, g \rangle$ *for all vectors* f *and* g *in* $\mathscr{H}$.

Proof. Let f and g be in $\mathscr{H}$ and write, using the projection theorem, $f = m_1 + n_1$, $g = m_2 + n_2$, where $m_1, m_2 \in M$ and $n_1, n_2 \in M^\perp$. We have

$$\langle f, Pg \rangle = \langle m_1 + n_1, m_2 \rangle = \langle m_1, m_2 \rangle$$

while

$$\langle Pf, g \rangle = \langle m_1, m_2 + n_2 \rangle = \langle m_1, m_2 \rangle.$$

$\square$

Returning to our question, if $f \in L^2(\mathbb{D}, dA/\pi)$, then for any $w \in \mathbb{D}$,

$$Pf(w) = \langle Pf, K_w \rangle = \langle f, PK_w \rangle = \langle f, K_w \rangle = \int_{\mathbb{D}} f(z) \overline{K_w(z)} \frac{dA}{\pi},$$

where K_w is the vector in $L_a^2(\mathbb{D})$ that gives the linear functional of evaluation at w, and we have used the lemma for the second equality. Since by Exercise 1.25 $K_w(z) = (1 - \overline{w}z)^{-2}$, this gives an integral formula for computing the projection Pf.

The Bergman space furnishes an example of what are called *functional Banach spaces*. Here is the definition: A Banach space X consisting of scalar-valued functions on a set S is a functional Banach space if point evaluation $e_s(f) \equiv f(s)$ at each point s of S is a bounded linear functional on X, and if no evaluation functional e_s is identically 0. Other examples of functional Banach spaces, besides $L_a^2(\mathbb{D})$, include $C[0,1]$ in the supremum norm and ℓ^p for $1 \le p \le \infty$. A non-example is $L^p([0,1], dx), 1 \le p \le \infty$; here the vectors are equivalence classes of functions, and evaluation at a point of $[0,1]$ doesn't even make sense.

1.5 Orthonormal Bases

Definition 1.31. An *orthonormal set* in a Hilbert space $\mathcal{H}$ is a set $\mathcal{E}$ with the properties:

(1)   for every $e \in \mathcal{E}$, $\|e\| = 1$, and
(2)   for distinct vectors e and f in $\mathcal{E}$, $\langle e, f \rangle = 0$.

For an easy example of an orthonormal set in the Hilbert space ℓ^2, take the set $\mathcal{E}$ of vectors $e_j, j \geq 1$ where e_j has a 1 in the jth coordinate and zeros elsewhere. As a second example, consider the Hilbert space $L^2[0, 2\pi]$, with respect to normalized Lebesgue measure $dt/(2\pi)$. The collection of functions e^{int} for any integer n form an orthonormal set in this Hilbert space. We often will write $L^2(T)$ for $L^2([0, 2\pi], dt/(2\pi))$, where T denotes the unit circle and we are identifying a function on $[0, 2\pi]$ with a function on T by $f(t) = f(e^{it})$.

Definition 1.32. An *orthonormal basis* for a Hilbert space $\mathcal{H}$ is a maximal orthonormal set; that is, an orthonormal set that is not properly contained in any orthonormal set.

It is easy to see that in the ℓ^2 example above, the set $\{e_j : j \geq 1\}$ is an orthonormal basis. Harder, but still true, is that $\{e^{int} : n \in \mathbb{Z}\}$, where $\mathbb{Z}$ is the set of all integers and $e^{int} = \cos(nt) + i\sin(nt)$, is an orthonormal basis for $L^2(T)$. This result is a consequence of Fejér's theorem; for a proof the reader is referred to [48]. Every Hilbert space *has* an orthonormal basis (see Exercise 3.1 in Chapter 3). The proof of this statement uses Zorn's lemma, which will be discussed in Section 3.1. The Hilbert spaces of principal interest to us will either have a finite or countably infinite orthonormal basis.

A Hilbert space is also a vector space, and as such it has a linear (or *Hamel*) basis. We digress here briefly to recall some facts and terminology from linear algebra. Given a nonempty subset S in a vector space V, by a *linear combination* of vectors in S we mean a *finite* sum of the form

$$\sum_{j=1}^{n} \alpha_j v_j$$

where the vectors v_j are in S and the coefficients α_j are scalars. A set S *spans* V if every vector in V is a (necessarily finite) linear combination of vectors in S. A set S of vectors is said to be linearly independent if the only linear combination of vectors in S that is equal to the zero vector is the one whose scalar coefficients are all zero. A linear, or Hamel, basis for a vector space V is a subset of V that is both linearly independent and spans V. It is easy to see that a Hamel basis can be equivalently defined as a maximal linearly independent subset of V; that is, a linearly independent set that is not properly contained in any linearly independent set.

Every vector space has a Hamel basis, and for a given vector space, any two Hamel bases can be put in one-to-one correspondence; proofs of these can be provided by the reader, or found, for example, in [14]. Notice that the concept of a

Hamel basis depends only on the linear structure of V, and not the topological structure that comes when the vector space is endowed with a norm or inner product. It is for this reason that in a Hilbert space the concept of an orthonormal basis proves to be more central than that of a Hamel basis, so much so that in the context of Hilbert spaces the term "basis" will always mean "orthonormal basis," and "dimension" will always refer to the (common) cardinality of any orthonormal basis. In particular, a Hilbert space is said to be finite-dimensional if it has a finite orthonormal basis, and infinite-dimensional otherwise. This convention will not lead to any confusion because of the following two facts: A finite orthonormal set in a Hilbert space $\mathcal{H}$ that is not properly contained in any orthonormal set is in fact a Hamel basis for $\mathcal{H}$, and no Hilbert space with a finite Hamel basis can contain an infinite orthonormal set. See Exercise 1.21 for a further exploration of these and related ideas.

Given a linearly independent sequence $\{f_n\}_1^\infty$ in a Hilbert space $\mathcal{H}$, there always exists an orthonormal sequence $\{e_n\}_1^\infty$ such that

$$\mathrm{span}\{f_1, f_2, \ldots, f_k\} = \mathrm{span}\{e_1, e_2, \ldots, e_k\}$$

for each positive integer k, where "span" denotes the set of linear combinations of the indicated set. An inductive process for constructing the vectors e_j, called *Gram–Schmidt orthonormalization*, is outlined in Exercise 1.19.

The last topic of this section is motivated by the question: When is an orthonormal set in a Hilbert space an orthonormal basis? When $\{e_k\}$ is a finite or countably infinite orthonormal *set* in $\mathcal{H}$, then for every vector $h \in \mathcal{H}$ we have

$$\sum |\langle h, e_k \rangle|^2 \leq \|h\|^2;$$

this is known as *Bessel's inequality*. It follows from the observation that the closest vector to h in the linear span of the orthonormal set $\{e_1, e_2, \ldots, e_n\}$ is $\sum_1^n \langle h, e_k \rangle e_k$ (see Exercise 1.21), and the Pythagorean identity of Proposition 1.22.

The identity in (e) of the next result is called *Parseval's identity*; it is the equality case of Bessel's inequality.

Theorem 1.33. *If $\{e_n\}_1^\infty$ is an orthonormal sequence in a Hilbert space $\mathcal{H}$, then the following conditions are equivalent:*

(a) $\{e_n\}_1^\infty$ *is an orthonormal basis.*
(b)   *If $h \in \mathcal{H}$ and $h \perp e_n$ for all n, then $h = 0$.*
(c)   *For every $h \in \mathcal{H}$, $h = \sum_1^\infty \langle h, e_n \rangle e_n$; equality here means the convergence in the norm of $\mathcal{H}$ of the partial sums to h.*
(d)   *For every $h \in \mathcal{H}$, there exist complex numbers a_n so that $h = \sum_1^\infty a_n e_n$.*
(e)   *For every $h \in \mathcal{H}$, $\sum_1^\infty |\langle h, e_n \rangle|^2 = \|h\|^2$.*
(f)   *For all h and g in $\mathcal{H}$, $\sum_1^\infty \langle h, e_n \rangle \langle e_n, g \rangle = \langle h, g \rangle$.*

Proof. The equivalence of (a) and (b) follows almost immediately from the definition, since if $0 \neq h$ and $h \perp e_n$ for all n, then $\{e_n\}_1^\infty \cup \{h/\|h\|\}$ is an orthonormal set.

Now assume (b) and suppose $h \in \mathcal{H}$ and let $c_n = \langle h, e_n \rangle$. By Bessel's inequality we have $\sum_1^\infty |c_n|^2 < \infty$, so that the partial sums $s_k = \sum_{n=1}^k c_n e_n$ form a Cauchy sequence in $\mathcal{H}$ with

$$\|s_k - s_m\|^2 = \sum_{n=m+1}^k |c_n|^2$$

whenever $k > m$. By completeness these partial sums must converge in $\mathcal{H}$ to some vector s. We claim $s = h$, which will show (c). For each fixed n,

$$\langle s, e_n \rangle = \langle \lim_{k \to \infty} s_k, e_n \rangle = \lim_{k \to \infty} \langle s_k, e_n \rangle = \langle h, e_n \rangle,$$

where we have used the continuity of the inner product, a consequence of the Cauchy–Schwarz inequality. Thus $\langle s - h, e_n \rangle = 0$ for all n, and by (b), $s = h$. The reverse implication, $(c) \Longrightarrow (b)$, is easy, since if (c) holds, and $h \perp e_n$ for all n, then $h = \sum_1^\infty \langle h, e_n \rangle e_n$ implies that $h = 0$.

Clearly (c) implies (d) and the reverse implication follows from setting $f_k = \sum_{j=1}^k a_j e_j$ and noting as above that

$$\langle h, e_n \rangle = \langle \lim_{k \to \infty} f_k, e_n \rangle = \lim_{k \to \infty} \langle f_k, e_n \rangle = a_n.$$

Next we show that (c) implies (e). Continuity of the norm shows that if $h = \sum_1^\infty \langle h, e_n \rangle e_n$, then $\|s_k\| \to \|h\|$ where s_k is the partial sum $\sum_{n=1}^k \langle h, e_n \rangle e_n$ and $\|s_k\|^2 = \sum_1^k |\langle h, e_n \rangle|^2$ by the Pythagorean formula. Thus (e) holds.

Clearly (f) implies (e) and the reverse implication can be obtained by using the polarization identity to write $\langle h, g \rangle$ in terms of $\|h + g\|^2, \|h - g\|^2, \|h + ig\|^2$ and $\|h - ig\|^2$, expanding each of these norms using (e), and computing.

Finally, if (e) holds, and $h \perp e_n$ for all n, then $\|h\|^2 = \sum_1^\infty |\langle h, e_n \rangle|^2 = 0$, giving (b). $\qquad\square$

When $\{e_n\}_1^\infty$ is an orthonormal basis for a Hilbert space $\mathcal{H}$, and $h \in \mathcal{H}$, the scalars $\langle h, e_n \rangle$ are called the *Fourier coefficients* of h with respect to $\{e_n\}_1^\infty$. In this case, the sum in (c) of the above theorem is referred to as the *Fourier series* of h, relative to the specified orthonormal basis.

No countably infinite orthonormal basis can ever be a Hamel basis. Indeed, using Gram–Schmidt orthonormalization we can show something stronger. Suppose that $\{f_1, f_2, \ldots\}$ is a linearly independent sequence in a Hilbert space $\mathcal{H}$. We claim that there is a vector in $\mathcal{H}$ which is not a finite linear combination of the f_j. The Gram–Schmidt process produces an orthonormal sequence $\{e_1, e_2, \ldots\}$ with span$\{e_1, e_2, \ldots, e_k\} = $ span$\{f_1, f_2, \ldots, f_k\}$ for all positive integers k. Write down any sum $\sum_1^\infty c_j e_j$ where infinitely many of the coefficients c_j are nonzero and $\sum_1^\infty |c_j|^2 < \infty$. This sum converges in $\mathcal{H}$ to some vector g, since its partial sums form a Cauchy sequence in $\mathcal{H}$. Since $\langle g, e_n \rangle = c_n$, and this is nonzero for infinitely many n, g is not in the span of $\{e_1, e_2, \ldots, e_k\}$ for any k, and hence neither is it in the span of $\{f_1, f_2, \ldots, f_k\}$ for any k.

Example 1.34. It is easy to see that if $e_n = \sqrt{n+1}\, z^n$, then $\{e_n\}_{n=0}^\infty$ is an orthonormal sequence in the Bergman space $L_a^2(\mathbb{D})$. Is it an orthonormal basis? It's

tempting to think we can make short work of answering this question. By part
(d) of Theorem 1.33, we need only show that every function f in the Bergman
space can be written as $f = \sum_0^\infty a_n e_n = \sum_0^\infty a_n \sqrt{n+1} z^n$, and since f is analytic in
the disk it has a power series expansion $f = \sum_0^\infty b_n z^n$. But the two equality signs, in
$f = \sum_0^\infty a_n \sqrt{n+1} z^n$ and in $f = \sum_0^\infty b_n z^n$, refer to two different kinds of convergence.
In the first, we want convergence of the partial sums in the $L^2(\mathbb{D}, dA/\pi)$ norm, while
the second gives us pointwise convergence in $\mathbb{D}$, or, better, uniform convergence on
compact subsets of $\mathbb{D}$ of the partial sums of $\sum_0^\infty b_n z^n$ to f. Since this latter type of
convergence does not imply L^2 convergence, something more must be done.

To that end, we will show that $\{e_n\}_{n=0}^\infty$ is an orthonormal basis for $L_a^2(\mathbb{D})$ by
showing that if $f \in L_a^2(\mathbb{D})$ and $f \perp e_n$ for all $n = 0, 1, 2 \ldots$, then $f = 0$. The assumption that $f \perp e_n$ is simply that

$$\int_{\mathbb{D}} f(z) \bar{z}^n \frac{dA}{\pi} = 0.$$

We can write f in terms of its power series in $\mathbb{D}$, $f(z) = \sum_0^\infty b_k z^k$, where the partial
sums of this series converge uniformly on compact subsets of $\mathbb{D}$. Fix $t < 1$ and use
this uniform convergence to write

$$\int_{t\mathbb{D}} f(z) \bar{z}^n \, dA = \int_{t\mathbb{D}} \left(\sum_{k=0}^\infty b_k z^k \right) \bar{z}^n \, dA$$

$$= \sum_{k=0}^\infty b_k \int_{t\mathbb{D}} z^k \bar{z}^n \, dA$$

$$= \pi \frac{b_n}{n+1} t^{2n+2}$$

since the integral in the penultimate line is 0 unless $k = n$, in which case it
is $\pi t^{2n+2}/(n+1)$. Now $f(z)\bar{z}^n$ is in $L^2(\mathbb{D}, dA/\pi) \subseteq L^1(\mathbb{D}, dA/\pi)$, so we can let
$t \uparrow 1$ and use the dominated convergence theorem to see that for any nonnegative
integer n,

$$\int_{\mathbb{D}} f(z) \bar{z}^n \frac{dA}{\pi} = \frac{b_n}{n+1}.$$

Thus if this integral is 0 for all n, we see that $b_n = 0$ for all n and hence $f = 0$ as
desired.

Specializing the result of (e) of Theorem 1.33 to $\mathscr{H} = L^2[a,b]$ we obtain the
Riesz–Fischer theorem, named for simultaneous and independent work of Riesz and
Ernst Fischer in 1907. More precisely, Fischer showed that $L^2[a,b]$ is complete (see
also Section A.3 in the Appendix), while Riesz showed that given a orthonormal
basis $\{e_n\}$ of $L^2[a,b]$, the map which sends $f \in L^2[a,b]$ to the square-summable
sequence $\{a_n\}$, defined by $a_n = \int_a^b f \bar{e}_n dx$, is an isometric linear bijection onto ℓ^2.
These two results are individually or collectively referred to as the "Riesz–Fischer
theorem"; they are equivalent in the sense that each can be recaptured from the other.
Theorem 1.29, the Riesz representation theorem (or as it should be more accurately

called, the Fréchet–Riesz Representation theorem) enters into the mix as well, in the following argument due to Riesz. Suppose we have an orthonormal basis $\{e_i\}$ (known to Riesz, based on a talk given by Erhard Schmidt it 1905, to be countable) for $L^2[a,b]$ and a sequence $\{c_i\}$ of complex numbers with $\sum |c_i|^2 < \infty$. Define Λ on $L^2[a,b]$ by

$$\Lambda(f) = \sum c_i \langle f, e_i \rangle,$$

where $\langle f, e_i \rangle = \int_a^b f \bar{e_i} dx$. The mapping Λ is linear, and bounded by the Cauchy–Schwarz inequality and Bessel's inequality. By Theorem 1.29, there exists $g \in L^2[a,b]$ such that

$$\int_a^b f \bar{g} dx = \Lambda(f) = \sum c_i \langle f, e_i \rangle$$

for all $f \in L^2[a,b]$. Setting $f = e_i$ gives $\int_a^b e_i \bar{g} dx = c_i$, and setting $f = g$ gives

$$\int_a^b |g|^2 dx = \sum |\langle g, e_i \rangle|^2.$$

Now imagine starting with $g \in L^2[a,b]$ and defining $c_i = \int_a^b g \bar{e_i} dx$. Defining Λ as above, we get the stated isometric linear bijection between $L^2[a,b]$ and ℓ^2. Riesz also provided three proofs for the completeness of $L^2[a,b]$. Exercise 1.27 outlines one of these, which relies on Theorem 1.29 as well.

1.6 Exercises

1.1. Prove the reverse triangle inequality: For vectors x, y in any normed linear space,

$$\|x+y\| \geq \|\|x\| - \|y\|\|.$$

1.2. Show that $C[0,1]$ is a Banach space in the supremum norm. Hint: If $\{f_n\}$ is a Cauchy sequence in $C[0,1]$, then for each fixed $x \in [0,1]$, $\{f_n(x)\}$ is a Cauchy sequence in $\mathbb{C}$, which is complete.

1.3. Let $C^1[0,1]$ be the space of continuous, complex-valued functions on $[0,1]$ with continuous first derivative. Show that in the supremum norm $\|\cdot\|_\infty$, $C^1[0,1]$ is not a Banach space, but that in the norm defined by $\|f\| = \|f\|_\infty + \|f'\|_\infty$ it does become a Banach space.

1.4. Show that the space ℓ^1 of Example 1.5 is complete.

1.5. Show that a metric space is complete if every Cauchy sequence has a convergent subsequence.

1.6. Assume that you know Minkowski's inequality

$$\left(\sum_{j=1}^n |a_j + b_j|^2 \right)^{1/2} \leq \left(\sum_{j=1}^n |a_j|^2 \right)^{1/2} + \left(\sum_{j=1}^n |b_j|^2 \right)^{1/2}$$

for $\mathbb{C}^n$ in the Euclidean norm.

(a) Show that for $\{a_n\}$ and $\{b_n\}$ in a weighted sequence space ℓ_β^2,

$$\left(\sum_{j=0}^\infty |a_j+b_j|^2 \beta(j)^2\right)^{1/2} \leq \left(\sum_{j=0}^\infty |a_j|^2 \beta(j)^2\right)^{1/2} + \left(\sum_{j=0}^\infty |b_j|^2 \beta(j)^2\right)^{1/2}.$$

(b) Verify directly (without appealing to the Riesz–Fischer theorem on the completeness of $L^2(X,\mu)$ in general) that ℓ_β^2 is complete.

1.7. Let x and y be any two vectors in an inner product space and set $\lambda = \langle y,y \rangle$. Show that

$$\lambda\left[\lambda\langle x,x\rangle - |\langle x,y\rangle|^2\right] = \langle \lambda x - \langle x,y\rangle y, \lambda x - \langle x,y\rangle y\rangle.$$

Use this to derive the Cauchy–Schwarz inequality and to determine when equality holds in the Cauchy–Schwarz inequality.

1.8. (a) Show that for a normed linear space X, the map $x \rightarrow \|x\|$ of X into $[0,\infty)$ is continuous. Is it uniformly continuous?

(b) Show that the mappings $X \times X \rightarrow X$ given by $(x,y) \rightarrow x+y$, and $\mathbb{C} \times X \rightarrow X$ given by $(\alpha,x) \rightarrow \alpha x$ are continuous. The topologies on $X \times X$ and $\mathbb{C} \times X$ are the product topologies.

(c) Suppose that X is an inner product space. Show that the maps $x \rightarrow \langle x,y\rangle$ and $x \rightarrow \langle y,x\rangle$ are continuous on X for each fixed y in X. Are they uniformly continuous?

1.9. (a) Show that if $\sum_0^\infty |a_n|^2 < \infty$, then the power series $\sum_0^\infty a_n z^n$ has radius of convergence at least one, and hence is an analytic function in the unit disk $\mathbb{D}$. Hint: Recall that the radius of convergence R is determined by

$$\frac{1}{R} = \limsup_{n\to\infty} |a_n|^{1/n}.$$

(b) Show that if

$$\lim_{n\to\infty} \beta(n)^{1/n} \geq 1$$

and

$$\sum_0^\infty |a_n|^2 \beta(n)^2 < \infty$$

then the power series

$$\sum_0^\infty a_n z^n$$

has radius of convergence at least equal to 1.

1.10. Suppose that X and Y are normed linear spaces and that $T : X \rightarrow Y$ is a linear map (meaning that $T(\alpha x_1 + \beta x_2) = \alpha T(x_1) + \beta T(x_2)$ for all vectors x_1, x_2 in X and all scalars α and β). Suppose that T maps X onto Y and is isometric (meaning $\|Tx\| = \|x\|$ for all $x \in X$).

(a) Show that T is one-to-one.
(b) Show that if X is a Banach space, so is Y.
(c) Show that if X is a Hilbert space, then so is Y if we define

$$\langle y_1, y_2 \rangle_Y = \langle x_1, x_2 \rangle_X$$

where x_1 and x_2 are the unique points in X satisfying $Tx_1 = y_1$ and $Tx_2 = y_2$.
(d) Explain how this shows that the weighted Hardy spaces $H^2(\beta)$ of Example 1.7 are Hilbert spaces.

1.11. For $1 \leq p < \infty$, define $L_a^p(\mathbb{D})$ to be the set of all analytic functions f on the unit disk $\mathbb{D}$ for which

$$\int_{\mathbb{D}} |f(z)|^p \frac{dA}{\pi} < \infty$$

and set $\|f\|^p$ to be the value of this integral. Show that $L_a^p(\mathbb{D})$ is a Banach space by first obtaining the appropriate analogue of Corollary 1.19.

1.12. Suppose S is a (not necessarily closed) subspace of a Hilbert space $\mathscr{H}$. Show that $S^{\perp\perp} \equiv (S^\perp)^\perp$ is the closure of S.

1.13. Show that $C[0, 1]$ in the supremum norm is not an inner product space; that is, the norm cannot be derived from an inner product.

1.14. Show that in any normed linear space where the norm satisfies the parallelogram equality, an inner product can be defined which induces the norm in the usual sense that $\langle x, x \rangle = \|x\|^2$. Hints: Define $\langle x, y \rangle$ by polarization and show that $\langle x, y \rangle = \overline{\langle y, x \rangle}$. Next show that $\langle x + y, z \rangle = \langle x, z \rangle + \langle y, z \rangle$ by showing the equality of the real parts and imaginary parts of both sides of this identity separately. Finally, show that $\langle sx, y \rangle = s\langle x, y \rangle$ for s in turn an integer, a rational number, a real number and a complex number.

1.15. Let M be a closed subspace of a Hilbert space $\mathscr{H}$, and suppose x_0 is in $\mathscr{H}$. Show that

$$\min\{\|m - x_0\| : m \in M\} = \max\{|\langle x_0, n \rangle| : n \in M^\perp, \|n\| = 1\}.$$

1.16. Let $\Lambda : X \rightarrow \mathbb{C}$ be a bounded linear functional on a normed linear space X. Recall that $\|\Lambda\|$ is defined as $\sup\{|\Lambda(x)| : \|x\| \leq 1\}$. Show that

$$\|\Lambda\| = \sup\{|\Lambda(x)| : \|x\| = 1\}$$
$$= \sup\{|\Lambda(x)|/\|x\| : x \neq 0\}$$
$$= \inf\{\delta : |\Lambda(x)| \leq \delta\|x\| \text{ for all } x \in X\}.$$

1.17. Let $1 < p < \infty$ and define q by $1/p + 1/q = 1$.

(a) Show that for each fixed $\{a_n\} \in \ell^q$, the linear mapping defined by

$$\Lambda(\{b_n\}) = \sum b_n \overline{a_n}$$

is a bounded linear functional on ℓ^p with norm $\|\{a_n\}\|_q$.

(b) Conversely, if Λ is a bounded linear functional on ℓ^p, then there exists $\{a_n\}$ in ℓ^q such that

$$\Lambda(\{b_n\}) = \sum b_n \overline{a_n}$$

for all $\{b_n\}$ in ℓ^p.

(c) What are the corresponding statements for the case $p = 1$?

1.18. Show that on the Hardy space H^2 as described in Example 1.7, evaluation at each point $w \in \mathbb{D}$ is a bounded linear functional. Hint: Use the Cauchy–Schwarz inequality to show that

$$|f(w)| \leq \|f\| \left(\frac{1}{1 - |w|^2} \right)^{1/2}.$$

1.19. In this problem we describe the Gram–Schmidt process: Let $x_1, x_2, \ldots$ be a sequence of linearly independent vectors in an inner product space. Define vectors inductively by

$$e_1 = x_1 / \|x_1\|$$

$$f_n = x_n - \sum_{j=1}^{n-1} \langle x_n, e_j \rangle e_j \text{ for } n \geq 2$$

$$e_n = f_n / \|f_n\| \text{ for } n \geq 2.$$

Show that $\{e_n\}$ is an orthonormal sequence with the property that the linear span of $\{x_1, x_2, \ldots, x_n\}$ is the same as the linear span of $\{e_1, e_2, \ldots, e_n\}$ for each n.

1.20. Apply the Gram–Schmidt process (Exercise 1.19) to the three vectors $\{1, x, x^2\}$ in $L^2([-1, 1], dx)$. Use your answer to find the distance from x^3 to the span of $\{1, x, x^2\}$; equivalently, find

$$\min_{a,b,c \in \mathbb{C}} \int_{-1}^{1} |x^3 - a - bx - cx^2|^2 dx.$$

When the Gram–Schmidt process is applied to the sequence $1, x, x^2, x^3, \ldots$, the resulting vectors are called the *Legendre polynomials*.

1.21. Let $\mathscr{H}$ be a Hilbert space.

(a) Every orthonormal set in $\mathscr{H}$ is linearly independent (recall this means that every finite subset is linearly independent).

(b) Suppose $\{e_1, e_2, \ldots, e_n\}$ is an orthonormal set in $\mathcal{H}$ and define

$$M \equiv \mathrm{span}\{e_1, e_2, \ldots, e_n\}.$$

Check that M is closed and show that if P is the projection of $\mathcal{H}$ onto M, then $Px = \sum_1^n \langle x, e_j \rangle e_j$ for all $x \in \mathcal{H}$.

(c) Show that if $\mathcal{H}$ has a finite orthonormal basis, it is also a Hamel basis for $\mathcal{H}$.

1.22. Suppose M is a closed subspace of a Hilbert space $\mathcal{H}$ and λ is a continuous linear functional on M with

$$\sup_{m \in M, \, m \neq 0} \frac{|\lambda(m)|}{\|m\|} = c.$$

Using Hilbert space methods, show that there is a unique continuous linear functional Λ on H with

$$\lambda(m) = \Lambda(m)$$

for all $m \in M$ and

$$\sup_{h \in H, \, h \neq 0} \frac{|\Lambda(h)|}{\|h\|} = c.$$

1.23. Let $\mathcal{H}$ be an infinite dimensional Hilbert space. Show that $\mathcal{H}$ has a countable orthonormal basis if and only if $\mathcal{H}$ has a countable dense subset.

1.24. Given a subset $E \subseteq \mathbb{N}$, consider the bounded sequence $x_E \equiv \{x_n\}_1^\infty$ with $x_n = 1$ if $n \in E$, and $x_n = 0$ otherwise.

(a) Show that for each $E \subseteq \mathbb{N}$, there is an open ball B_E in ℓ^∞ centered at x_E such that for distinct subsets E and F of $\mathbb{N}$, B_E and B_F are disjoint.

(b) Conclude that ℓ^∞ contains no countable dense subset. This says that ℓ^∞ is non-separable.

1.25. (a) Show that if $f \in L_a^2(\mathbb{D})$ has Taylor series expansion $\sum_0^\infty a_n z^n$ in $\mathbb{D}$, then

$$\|f\|^2 = \sum_{n=0}^\infty |a_n|^2 \frac{1}{n+1}.$$

Notice that this says that $L_a^2(\mathbb{D})$ is a weighted Hardy space $H^2(\beta)$ for $\beta(n) = (n+1)^{-\frac{1}{2}}$. Also, find an expression in terms of the Taylor coefficients of f and g for $\langle f, g \rangle_{L_a^2(\mathbb{D})}$, where f and g are in $L_a^2(\mathbb{D})$.

(b) We have seen that evaluation at $w \in \mathbb{D}$ is a bounded linear functional on $L_a^2(\mathbb{D})$. Thus there is a function, call it $K_w(z)$, in $L_a^2(\mathbb{D})$ so that $f(w) = \langle f, K_w \rangle$ for all $f \in L_a^2(\mathbb{D})$. Show that $K_w(z) = (1 - \bar{w}z)^{-2}$. What is the norm of the linear functional of evaluation at w?

1.26. Show that if $f(z) = \sum_{n=0}^\infty a_n z^n$ is in the Hardy space H^2, then for each $0 < r < 1$

$$\int_0^{2\pi} |f_r(e^{i\theta})|^2 \frac{d\theta}{2\pi} = \sum_{n=0}^{\infty} |a_n|^2 r^{2n},$$

where $f_r(e^{i\theta}) = f(re^{i\theta})$. Conclude that

$$\lim_{r \to 1^-} \int_0^{2\pi} |f(re^{i\theta})|^2 \frac{d\theta}{2\pi}$$

is equal to the H^2 norm of f. Conversely, show that if $f(z) = \sum_{n=0}^{\infty} a_n z^n$ is analytic in the unit disk $\mathbb{D}$ and

$$\lim_{r \to 1^-} \int_0^{2\pi} |f(re^{i\theta})|^2 \frac{d\theta}{2\pi} < \infty$$

then $\sum_{n=0}^{\infty} |a_n|^2 < \infty$.

1.27. [21] Complete the argument outlined below, due to Riesz, to give a proof of the completeness of $L^2[a,b]$. Let $\{f_n\}$ be a Cauchy sequence in $L^2[a,b]$.

(a) Show that $\{\|f_n\|\}$ is a Cauchy sequence of scalars, and hence a bounded sequence, say $\|f_n\| \le M$.
(b) Show that for fixed $g \in L^2[a,b]$, $\int_a^b f_n g \, dx$ is a Cauchy sequence and its limit α_g satisfies $|\alpha_g| \le M\|g\|$.
(c) Define $\Lambda : L^2[a,b] \to \mathbb{C}$ by $\Lambda(g) = \alpha_g$. Use Theorem 1.29 to show that there exists $F \in L^2[a,b]$ with $\alpha_g = \int_a^b F g \, dx$ for all $g \in L^2[a,b]$.
(d) Show that f_n converges to F in $L^2[a,b]$.

1.28. A sequences $\{h_n\}$ in a Hilbert space $\mathscr{H}$ is said to *converge weakly* to $h \in \mathscr{H}$ if

$$\lim_{n \to \infty} \langle h_n, g \rangle = \langle h, g \rangle$$

for every $g \in \mathscr{H}$.

(a) If $\{e_n\}$ is an orthonormal sequence in $\mathscr{H}$, show that $e_n \to 0$ weakly.
(b) Show that if $h_n \to h$ in norm, then $h_n \to h$ weakly. Show that the converse is false, but that if $h_n \to h$ weakly and $\|h_n\| \to \|h\|$, then $h_n \to h$ in norm.

1.29. Show that if $\{f_n\}$ is a sequence in $L_a^2(\mathbb{D})$ and $f_n \to f$ weakly in $L_a^2(\mathbb{D})$ (see the previous exercise for the definition), then $f_n(z) \to f(z)$ for each $z \in \mathbb{D}$.

1.30. Let $0 < p < 1$ and define $L^p[0,1]$, with respect to Lebesgue measure dx, to be the set of all (equivalence classes of) measurable functions for which

$$\int_0^1 |f|^p dx < \infty.$$

(a) Show that if we define

$$d(f,g) = \int_0^1 |f - g|^p dx$$

for $f, g \in L^p[0,1]$, then d is a metric on $L^p[0,1]$ and the resulting metric space is complete.

(b) Show that $\|\cdot\|_p$ does not satisfy the triangle inequality and thus is not a norm on $L^p[0,1]$, where as usual we write

$$\|f\|_p = \left(\int_0^1 |f|^p dx \right)^{1/p}.$$

1.31. Suppose that $\mathscr{H}_1, \mathscr{H}_2, \ldots$ is a finite or countable collection of Hilbert spaces. The purpose of this exercise is to define the (external) direct sum of the spaces $\mathscr{H}_n$. We give the definition in the case of a countable collection and leave it to the reader to describe the obvious modifications in the finite case. Define $\mathscr{H}$ to be the set of all sequences $\{h_n\}$ with $h_n \in \mathscr{H}_n$ for each n and $\sum_1^\infty \|h_n\|^2 < \infty$. Addition and scalar multiplication on $\mathscr{H}$ are defined coordinatewise, and for $h = \{h_n\}$ and $g = \{g_n\}$ in $\mathscr{H}$ we define

$$\langle h, g \rangle_{\mathscr{H}} = \sum_{n=1}^\infty \langle h_n, g_n \rangle_{\mathscr{H}_n}.$$

(a) Show that $\langle \cdot, \cdot \rangle$ is an inner product on $\mathscr{H}$, and in the resulting norm $\|h\|^2 = \sum_1^\infty \|h_n\|^2$, $\mathscr{H}$ is a Hilbert space. (Use the Cauchy–Schwarz inequality to show that the sum in the definition of $\langle h, g \rangle_{\mathscr{H}}$ converges absolutely.)

(b) Denote $\mathscr{H}$ as just defined by $\sum \oplus \mathscr{H}_n$. Sometimes we like to think of $\mathscr{H}_n$ as a subspace in $\sum \oplus \mathscr{H}_n$; this means that we identify $\mathscr{H}_n$ with those elements of $\sum \oplus \mathscr{H}_n$ which have a zero in all but the nth position. Show that with this identification, $\mathscr{H}_n \perp \mathscr{H}_m$ for $m \neq n$.

1.32. In this problem we compare the results of the previous exercise with a slightly different, but essentially equivalent, notion of the direct sum of Hilbert spaces. Suppose that $\mathscr{H}_1, \mathscr{H}_2, \ldots$ is a collection of pairwise orthogonal closed subspaces of a Hilbert space $\mathscr{H}$. By $\sum \oplus \mathscr{H}_n$ (which we temporarily call the internal direct sum of the spaces $\mathscr{H}_n$) we mean the closure of the collection of all finite sums $h_1 + h_2 + \cdots + h_m$ where $m \in \mathbb{N}$ and $h_j \in \mathscr{H}_j$ for all j. Show that this internal direct sum is isomorphic to the external direct sum via the correspondence

$$(h_1, h_2, h_3, \ldots) \leftrightarrow h_1 + h_2 + h_3 + \cdots.$$

1.33. If $\mathscr{H}_1, \mathscr{H}_2, \ldots$ is a collection of pairwise orthogonal closed subspaces of a Hilbert space $\mathscr{H}$, and $\sum \oplus \mathscr{H}_n$ is their internal direct sum (in the terminology of the previous exercise), show that $\sum \oplus \mathscr{H}_n$ is the intersection of all closed subspaces containing $\cup_n \mathscr{H}_n$. This is called the *closed linear span* of the subspaces $\mathscr{H}_n$ and is denoted $\bigvee_n \mathscr{H}_n$. Thus we have

$$\bigvee_n \mathscr{H}_n = \sum_n \oplus \mathscr{H}_n$$

for pairwise orthogonal $\mathscr{H}_n$.

Chapter 2
Operator Theory Basics

The constantly widening field of applications of functional analysis leads to a systematic reconsideration of its basic methodological standpoints. One of these standpoints asserts that the original and basic concept of functional analysis is the concept of a space (normed, metric,...). To study a problem one must choose a space and study the corresponding functionals, operators, etc. in it. ... [T]he choice of the space in which the problem is studied is partly connected with the subjective aims which the investigator sets himself. Apparently the objective data are only the operators that appear in the equations of the problem. On this account it seems to us that the original and basic concept of functional analysis is that of an operator.
S. Krein and Yu. Petunin ([28], p. 85).

Linear operators connect, either explicitly or in the background, to all of the topics of this book, so in this chapter we will discuss the most basic properties of operators on Banach or Hilbert spaces.

2.1 Bounded Linear Operators

Definition 2.1. If X and Y are normed linear spaces, a map $T : X \to Y$ is *linear* if

$$T(\alpha x_1 + \beta x_2) = \alpha(Tx_1) + \beta T(x_2)$$

for all x_1, x_2 in X and scalars α and β. We say the linear map T is a *bounded linear operator* from X to Y if there is a finite constant C such that $\|Tx\|_Y \leq C\|x\|_X$ for all x in X.

We will normally suppress the subscript on the norm symbol $\|\cdot\|$, which indicates the space in which the vector lives, unless there is a potential for confusion. Bounded linear functionals introduced in the last chapter are bounded linear operators for the special case $Y = \mathbb{C}$. As with linear functionals, boundedness of a linear operator is equivalent to continuity, as the next result, whose proof is left to the reader, states.

Proposition 2.2. *If $T : X \to Y$ is a linear map from a normed linear space X to a normed linear space Y, the following are equivalent:*

(a) T *is bounded.*
(b) T *is continuous.*
(c) T *is continuous at 0.*

B.D. MacCluer, *Elementary Functional Analysis*, DOI 10.1007/978-0-387-85529-5_2,
© Springer Science+Business Media, LLC 2009

As with linear functionals, we define $\|T\| = \sup\{\|Tx\| : \|x\| \le 1\}$ and refer to this as the "operator norm of T." This terminology is justified by the following result, whose proof is left as Exercise 2.1.

Proposition 2.3. *The collection $\mathscr{B}(X,Y)$ of all bounded linear operators from a normed linear space X to a normed linear space Y is a normed linear space in the operator norm, where the vector operations are defined pointwise. If, in addition, Y is a Banach space, then $\mathscr{B}(X,Y)$ is a Banach space.*

When $X = Y$ we will write $\mathscr{B}(X)$ for $\mathscr{B}(X,X)$. Note that, as promised in Section 1.4, Proposition 2.3 tells us that the collection of all bounded linear functionals on a normed linear space forms a Banach space, since $Y = \mathbb{C}$ is complete.

Before discussing any more of the general theory of bounded linear operators it's helpful to have a list of examples in mind.

Example 2.4. Here is an example we have already met. Suppose M is a closed subspace in a Hilbert space $\mathscr{H}$. Let $P_M : \mathscr{H} \to M$ be the orthogonal projection of $\mathscr{H}$ onto M. By the projection theorem, P_M is a bounded linear operator of norm 1 (if $M \ne \{0\}$).

Example 2.5. Let $(X, \mathfrak{M}, \mu)$ be any σ-finite measure space and choose $\varphi \in L^\infty(X, \mu)$. Define the *multiplication operator* $M_\varphi : L^2(X, \mu) \to L^2(X, \mu)$ by $M_\varphi(f) = \varphi f$; this is clearly a linear map and since

$$\int_X |\varphi f|^2 d\mu \le \|\varphi\|_\infty^2 \int_X |f|^2 d\mu$$

we see that M_φ is bounded with norm at most $\|\varphi\|_\infty$. We claim that in fact $\|M_\varphi\| = \|\varphi\|_\infty$. To see this suppose $\alpha < \|\varphi\|_\infty$. If E is defined to be $\{x : |\varphi(x)| > \alpha\}$, then $\mu(E) > 0$. The idea is to consider something like χ_E to show that $\|M_\varphi\| > \alpha$, but we can't quite do this since χ_E will not be in $L^2(X, \mu)$ if $\mu(E)$ is infinite. Instead, we use the σ-finiteness hypothesis on μ to find a subset E' of E with $0 < \mu(E') < \infty$. Then $f = \chi_{E'}$ will be in $L^2(X, \mu)$ and

$$\|M_\varphi f\|^2 = \int_{E'} |\varphi|^2 d\mu > \alpha^2 \mu(E') = \alpha^2 \int_X |f|^2 d\mu$$

so that $\|M_\varphi\| > \alpha$, where α is any chosen value less that $\|\varphi\|_\infty$. Thus $\|M_\varphi\| = \|\varphi\|_\infty$.

In the next example, which combines the previous two, we encounter the composition of two linear operators. In general, if $A \in \mathscr{B}(X,Y)$ and $B \in \mathscr{B}(Y,V)$, then by the product BA we mean the mapping defined by $BA(x) = B(A(x))$ for $x \in X$. It is easy to see that $BA \in \mathscr{B}(X,V)$ and $\|BA\| \le \|B\|\|A\|$.

Example 2.6. Consider the Bergman space $L_a^2(\mathbb{D})$ and let $\varphi \in L^\infty(\mathbb{D}, dA/\pi)$; note that φ is not assumed to be analytic. Define the *Toeplitz operator* with symbol φ on $L_a^2(\mathbb{D})$ by $T_\varphi(f) = P(\varphi f)$, where P is the projection of $L^2(\mathbb{D}, dA/\pi)$ onto $L_a^2(\mathbb{D})$.

Clearly T_φ is linear, and since $P : L^2(\mathbb{D}, dA/\pi) \to L^2_a(\mathbb{D})$ is bounded with norm 1, and $M_\varphi : L^2(\mathbb{D}, dA/\pi) \to L^2(\mathbb{D}, dA/\pi)$ is bounded of norm $\|\varphi\|_\infty$, T_φ is bounded with norm at most $\|\varphi\|_\infty$. When φ is analytic in $L^\infty(\mathbb{D}, dA/\pi)$, the projection factor in the definition of T_φ serves no purpose, and in this case T_φ is just the restriction of the multiplication operator M_φ to the closed subspace $L^2_a(\mathbb{D})$ of $L^2(\mathbb{D})$.

Example 2.7. The next pair of operators are simple but important ones. They act from ℓ^2 to itself. The first, called the *forward shift*, is defined by

$$S(x_1, x_2, \ldots) = (0, x_1, x_2, \ldots).$$

It is easy to see that it is a bounded linear operator of norm one; in fact it is an isometry, meaning $\|Sx\| = \|x\|$ for every $x = (x_1, x_2, \ldots) \in \ell^2$. The *backward shift* is the operator from ℓ^2 to ℓ^2 which takes $(x_1, x_2, x_3, \ldots)$ to $(x_2, x_3, \ldots)$. It has norm 1, but is not an isometry (why?).

Example 2.8. Suppose that $\mathcal{H}$ is a Hilbert space with orthonormal basis $\{e_n\}_1^\infty$. Choose any *bounded* sequence of complex numbers $\{\alpha_n\}_1^\infty$ and set $Ae_n = \alpha_n e_n$. Extend A by linearity to any finite linear combination of the e_n, and extend A to all of $\mathcal{H}$ by continuity, noting that linearity is preserved. Explicitly this means the following. Given $h \in \mathcal{H}$, we know $h = \sum_1^\infty \langle h, e_n \rangle e_n$, where the sum converges in $\mathcal{H}$. Since $\{\alpha_n\}$ is a bounded sequence, the partial sums of $\sum_1^\infty \langle h, e_n \rangle \alpha_n e_n$ form a Cauchy sequence in $\mathcal{H}$ and thus converge in $\mathcal{H}$; call the sum Ah. Since

$$\|Ah\|^2 = \sum_1^\infty |\langle h, e_n \rangle|^2 |\alpha_n|^2 \le (\sup_n |\alpha_n|^2) \sum_1^\infty |\langle h, e_n \rangle|^2 = (\sup_n |\alpha_n|)^2 \|h\|^2,$$

we see that A is bounded with $\|A\| \le \sup_n |\alpha_n|$. Consideration of Ae_n shows that, indeed, we have equality here. Such an operator A is called a *diagonal operator*, with diagonal sequence $\{\alpha_n\}$. The terminology comes from defining, in analogy with the finite-dimensional case, the matrix M_A of A (with respect to the basis $\{e_n\}$) to be the (infinite) matrix with ijth entry $\langle Ae_j, e_i \rangle$. This is a diagonal matrix when A is a diagonal operator.

In spite of its usefulness in finite-dimensional settings, we will not find it particularly helpful to work with the matrix of a general bounded linear operator on a Hilbert space $\mathcal{H}$, in part because it is not easy to tell if a linear operator A is bounded by looking at its matrix M_A. Another way to say this is that while every bounded linear operator corresponds to a matrix, the converse is not true.

Example 2.9. We describe a class of operators called *integral operators*. Start with a σ-finite measure space $(X, \mathfrak{M}, \mu)$ and a measurable function $k : X \times X \to \mathbb{C}$ with $k \in L^2(X \times X, \mu \times \mu)$. Define $K : L^2(X, \mu) \to L^2(X, \mu)$ by $Kf = g$ where

$$g(x) = \int_X k(x, y) f(y) d\mu(y)$$

for x in X. We call k the kernel of the integral operator K. Since $k \in L^2(X \times X)$, for almost every $x \in X$, the function $y \mapsto k(x, y)$ is in $L^2(X, \mu)$, and thus the function

$y \mapsto k(x,y)f(y)$ is integrable if $f \in L^2(X,\mu)$. We show that K maps into $L^2(X,\mu)$ and is bounded. For $f \in L^2(X,\mu)$ we have

$$
\begin{aligned}
\|K(f)\|^2 &= \int_X |g(x)|^2 d\mu(x) = \int_X \left| \int_X k(x,y)f(y)d\mu(y) \right|^2 d\mu(x) \\
&\leq \int_X \left(\int_X |k(x,y)||f(y)|d\mu(y) \right)^2 d\mu(x) \\
&\leq \int_X \left(\int_X |k(x,y)|^2 d\mu(y) \right) \left(\int_X |f(y)|^2 d\mu(y) \right) d\mu(x) \\
&= \|f\|^2 \int_X \int_X |k(x,y)|^2 d\mu(y)d\mu(x) \\
&= \|f\|^2 \|k\|^2,
\end{aligned}
$$

where we have used the Cauchy–Schwarz inequality midway through the calculation. This computation shows that K is bounded from $L^2(X,\mu)$ into $L^2(X,\mu)$ with norm at most $\|k\|$. Some of the measure-theoretic technicalities of this example can be bypassed by taking, for example, $X = [0,1]$ and requiring $k(x,y)$ to be continuous on $[0,1] \times [0,1]$.

A particular integral operator of interest is the Volterra operator; it comes from the choice $X = [0,1]$, with Lebesgue measure, and $k(x,y)$ equal to the characteristic function of the lower triangle $\{(x,y) : y \leq x\}$ in the unit square $[0,1] \times [0,1]$. This gives

$$
Kf(x) = \int_0^x f(y)dy,
$$

so the Volterra operator is sometimes called the "operator of indefinite integration." By the above remarks, its norm is at most $1/\sqrt{2}$; computing the norm exactly is not so easy; see [17], Problem 188.

A bounded linear operator T on a Banach space X is said to attain its norm if there is a nonzero vector x in X with $\|Tx\| = \|T\|\|x\|$. See Exercise 2.8 for an exploration of this issue in the Hilbert space setting.

2.2 Adjoints of Hilbert Space Operators

Now that we have some examples of bounded linear operators in mind, let us turn to the notion of the adjoint of a Hilbert space operator. Later we will define adjoints of operators on Banach spaces, and compare it to the definition we give now in the Hilbert space setting. As motivation for our work in this section, recall that given an $n \times n$ matrix $A = (a_{ij})$ with complex entries, its conjugate transpose A^* is the $n \times n$ matrix whose ijth entry is $\overline{a_{ji}}$. Associate to the matrix A the linear operator T_A on $\mathbb{C}^n$ given by $T_A(v) = Av$ where $v \in \mathbb{C}^n$ is written as a column vector. For any vectors v and w in $\mathbb{C}^n$, we have

$$\langle T_A v, w \rangle = \langle v, T_{A^*} w \rangle. \tag{2.1}$$

The operator T_{A^*} is called the adjoint of the operator T_A, and the analogue of the property in Equation (2.1) will lead us to the idea of the adjoint of any bounded linear operator between Hilbert spaces. To make this precise, we begin with the definition of a sesquilinear form.

Definition 2.10. If $\mathcal{H}$ and $\mathcal{K}$ are both Hilbert spaces, a *sesquilinear form* $u : \mathcal{H} \times \mathcal{K} \to \mathbb{C}$ is a mapping satisfying

(1) $u(\alpha h + \beta g, k) = \alpha u(h,k) + \beta u(g,k)$, and
(2) $u(h, \alpha k + \beta f) = \overline{\alpha} u(h,k) + \overline{\beta} u(h,f)$

for all $h, g \in \mathcal{H}$, all $k, f \in \mathcal{K}$ and all scalars α and β. A sesquilinear form u is bounded if there is a finite constant M such that $|u(h,k)| \leq M \|h\| \|k\|$ for all $h \in \mathcal{H}$ and $k \in \mathcal{K}$.

Setting $\mathcal{H} = \mathcal{K}$ and letting $u(h,k)$ be the inner product $\langle h, k \rangle$ gives an example of a sesquilinear form that is bounded (by the Cauchy–Schwarz inequality). More generally, if $A \in \mathcal{B}(\mathcal{H}, \mathcal{K})$ and $B \in \mathcal{B}(\mathcal{K}, \mathcal{H})$, then both $u(h,k) \equiv \langle Ah, k \rangle$ and $u(h,k) \equiv \langle h, Bk \rangle$ (where the inner products are in $\mathcal{K}$ and $\mathcal{H}$, respectively) define sesquilinear forms that are bounded, since, for example,

$$|u(h,k)| = |\langle Ah, k \rangle| \leq \|Ah\| \|k\| \leq \|A\| \|h\| \|k\|.$$

The next result describes all bounded sesquilinear forms.

Theorem 2.11. *Let $\mathcal{H}$ and $\mathcal{K}$ be Hilbert spaces and suppose that $u : \mathcal{H} \times \mathcal{K} \to \mathbb{C}$ is a bounded sesquilinear form. There exists a unique $A \in \mathcal{B}(\mathcal{H}, \mathcal{K})$ such that*

$$u(h,k) = \langle Ah, k \rangle_{\mathcal{K}}$$

for all $h \in \mathcal{H}$ and $k \in \mathcal{K}$.

Proof. For fixed $h \in \mathcal{H}$ we define a mapping $\Lambda_h : \mathcal{K} \to \mathbb{C}$ by

$$\Lambda_h(k) = \overline{u(h,k)}.$$

One can easily check that Λ_h is linear. Moreover, since u is bounded by hypothesis,

$$|\Lambda_h(k)| = |\overline{u(h,k)}| = |u(h,k)| \leq M \|h\| \|k\|$$

for some M independent of h and k. Thus Λ_h is a bounded linear functional on $\mathcal{K}$. By the Riesz representation theorem this functional must therefore be given by inner product with a unique vector f in $\mathcal{K}$:

$$\Lambda_h(k) = \langle k, f \rangle_{\mathcal{K}}$$

for all $k \in \mathcal{K}$, and moreover $\|\Lambda_h\| = \|f\|$. Since we have already observed that $\|\Lambda_h\| \leq M \|h\|$, we must have $\|f\| \leq M \|h\|$. This process defines a map A from $\mathcal{H}$

to $\mathscr{K}$ taking h to f. This mapping is linear: If h_1, h_2 are in $\mathscr{H}$ and $\alpha \in \mathbb{C}$ with $Ah_1 = f_1$ and $Ah_2 = f_2$ so that

$$\overline{u(h_1, k)} = \langle k, f_1 \rangle_{\mathscr{K}}$$

and

$$\overline{u(h_2, k)} = \langle k, f_2 \rangle_{\mathscr{K}}$$

for all $k \in \mathscr{K}$, then

$$\Lambda_{\alpha h_1 + h_2}(k) = \overline{u(\alpha h_1 + h_2, k)} = \overline{\alpha u(h_1, k)} + \overline{u(h_2, k)}$$
$$= \overline{\alpha} \langle k, f_1 \rangle_{\mathscr{K}} + \langle k, f_2 \rangle_{\mathscr{K}} = \langle k, \alpha f_1 + f_2 \rangle_{\mathscr{K}},$$

so that A maps $\alpha h_1 + h_2$ to $\alpha f_1 + f_2$. We have already seen that A is bounded with $\|Ah\| = \|f\| \le M\|h\|$. This shows that there is a bounded linear operator A with

$$\overline{u(h, k)} = \langle k, Ah \rangle_{\mathscr{K}}$$

or equivalently

$$u(h, k) = \langle Ah, k \rangle_{\mathscr{K}}.$$

Moreover, A is unique, since if $\langle f_1, k \rangle_{\mathscr{K}} = \langle f_2, k \rangle_{\mathscr{K}}$ for all $k \in \mathscr{K}$, we must have $\langle f_1 - f_2, k \rangle_{\mathscr{K}} = 0$ for all k, and therefore $f_1 = f_2$.   □

As a consequence of the last result, suppose we start with an operator A in $\mathscr{B}(\mathscr{H}, \mathscr{K})$ and define $u : \mathscr{K} \times \mathscr{H} \to \mathbb{C}$ by

$$u(k, h) \equiv \langle k, Ah \rangle_{\mathscr{K}}.$$

This is a bounded sesquilinear form. Applying Theorem 2.11, we can find the unique operator, call it A^*, in $\mathscr{B}(\mathscr{K}, \mathscr{H})$ satisfying

$$u(k, h) = \langle A^* k, h \rangle_{\mathscr{H}}$$

for all $k \in \mathscr{K}$ and $h \in \mathscr{H}$. Taking conjugates, we have the following important conclusion.

Theorem 2.12. *Given Hilbert spaces $\mathscr{H}$ and $\mathscr{K}$ and $A \in \mathscr{B}(\mathscr{H}, \mathscr{K})$, there is a unique $A^* \in \mathscr{B}(\mathscr{K}, \mathscr{H})$ so that*

$$\langle Ah, k \rangle_{\mathscr{K}} = \langle h, A^* k \rangle_{\mathscr{H}}$$

for all $h \in \mathscr{H}$ and $k \in \mathscr{K}$.

The operator A^* in the last result is called the (Hilbert space) adjoint of A. In the case that $\mathscr{H} = \mathscr{K}$ and $A^* = A$ we say that A is *self-adjoint* or *Hermitian*. Looking back at the statement of Lemma 1.30 in Chapter 1, we see that orthogonal projections (onto closed subspaces) are self-adjoint operators.

Some more examples are in order. For the forward shift S on ℓ^2 as in Example 2.7, it is easy to check that S^* is the backward shift; to see this it suffices to compute

$\langle Sx, y \rangle$ and $\langle x, By \rangle$ and see that they agree, where x and y are in ℓ^2 and B denotes the backward shift.

For a multiplication operator M_φ, defined on $L^2(X, \mu)$ for some σ-finite measure space (X, μ) and $\varphi \in L^\infty(X, \mu)$, we have

$$\langle M_\varphi f, g \rangle = \langle \varphi f, g \rangle = \int_X \varphi f \bar{g} d\mu = \langle f, \bar{\varphi} g \rangle = \langle f, M_{\bar{\varphi}} g \rangle$$

for any $f, g \in L^2(X, \mu)$. Thus $M_\varphi^* = M_{\bar{\varphi}}$, and a multiplication operator is self-adjoint if and only if its symbol φ is real-valued almost everywhere.

A similar computation, using the self-adjointness of the Bergman projection from Lemma 1.30, shows that a Toeplitz operator T_φ on $L_a^2(\mathbb{D})$ has adjoint $T_{\bar{\varphi}}$. See Exercise 2.5.

To determine the adjoint of the integral operator with kernel k acting on $L^2(X, \mu)$ we seek the operator K^* satisfying $\langle Kf, g \rangle = \langle f, K^*g \rangle$ for all f and g in $L^2(X, \mu)$. Writing the inner product as an integral, and using Fubini's theorem to interchange the order of integration, we see that

$$\begin{aligned}
\langle Kf, g \rangle &= \int_X Kf(x)\overline{g(x)}d\mu(x) \\
&= \int_X \left(\int_X k(x, y)\overline{g(x)}d\mu(x) \right) f(y)d\mu(y) \\
&= \int_X \left(\overline{\int_X \overline{k(x, y)}g(x)d\mu(x)} \right) f(y)d\mu(y),
\end{aligned}$$

which is equal to

$$\int_X f(y)\overline{K^*g(y)}d\mu(y) = \langle f, K^*g \rangle$$

if we define K^* to be the integral operator with kernel $k^*(x, y) \equiv \overline{k(y, x)}$.

We will be mainly interested in adjoints for bounded linear operators from a Hilbert space $\mathcal{H}$ to itself; recall we will write $\mathcal{B}(\mathcal{H})$ for $\mathcal{B}(\mathcal{H}, \mathcal{H})$ in this case. For simplicity, the next result is stated in this setting, rather than the more general one of bounded linear operators from $\mathcal{H}$ to $\mathcal{K}$.

Proposition 2.13. *For A and B in $\mathcal{B}(\mathcal{H})$ we have*

(a) $A^{**} = A$ where $A^{**} = (A^*)^*$.
(b) $(A + B)^* = A^* + B^*$.
(c) $(\alpha A)^* = \bar{\alpha}A^*$ for $\alpha \in \mathbb{C}$.
(d) $(AB)^* = B^*A^*$.

Proof. For (a) we first note that by the definition of the adjoint we have $\langle A^*x, y \rangle = \langle x, A^{**}y \rangle$ for all x and y in $\mathcal{H}$. Since also $\langle Ay, x \rangle = \langle y, A^*x \rangle$, taking conjugates we see that $\langle x, Ay \rangle = \langle A^*x, y \rangle$. Thus $\langle x, A^{**}y \rangle = \langle x, Ay \rangle$ for all x and y, or equivalently $\langle x, A^{**}y - Ay \rangle = 0$ for all x and y. This forces $A^{**}y = Ay$ for all $y \in \mathcal{H}$, giving (a).

Part (b) follows from the definition of the adjoint and a straightforward computation showing that $\langle (A+B)x, y \rangle = \langle x, (A^* + B^*)y \rangle$ for all x and y in $\mathscr{H}$. Parts (c) and (d) are done similarly, and the details are left to the reader.   □

The next result will be fundamentally important to us.

Proposition 2.14. *If $A \in \mathscr{B}(\mathscr{H})$, then $\|A\| = \|A^*\|$ and $\|A^*A\| = \|A\|^2$.*

Proof. Take any vector $h \in \mathscr{H}$ with $\|h\| = 1$. We have

$$\|Ah\|^2 = \langle Ah, Ah \rangle = \langle h, A^*Ah \rangle \leq \|A^*Ah\|\,\|h\| \leq \|A^*A\| \leq \|A^*\| \cdot \|A\| \qquad (2.2)$$

so that $\|A\| \leq \|A^*\|$. Applying this together with (a) of the previous proposition we also have $\|A^*\| \leq \|A^{**}\| = \|A\|$. Thus $\|A\| = \|A^*\|$. Using this, and taking the supremum over all unit vectors h in (2.2), we see that

$$\|A\|^2 = \sup\{\|Ah\|^2 : \|h\| = 1\} \leq \|A^*A\| \leq \|A^*\| \cdot \|A\| = \|A\|^2$$

and thus equality must hold throughout, yielding $\|A\|^2 = \|A^*A\|$ as desired.   □

Let us recap and extend the structure we have on $\mathscr{B}(\mathscr{H})$ when $\mathscr{H}$ is any Hilbert space. First of all, using pointwise-defined vector operations and the "operator norm," it is a Banach space; this is Proposition 2.3. We define a multiplication on $\mathscr{B}(\mathscr{H})$ which makes it into a *complex algebra*; that is, a vector space over $\mathbb{C}$ with a multiplication satisfying $A(BC) = (AB)C$, $(A+B)C = AC + BC$, $A(B+C) = AB + AC$, and $\alpha(AB) = (\alpha A)B = A(\alpha B)$ for all $A, B, C \in \mathscr{B}(\mathscr{H})$ and scalar α. This multiplication AB is just the composition of the linear maps A and B. Note multiplication is not in general commutative. As we have noted above, $\|AB\| \leq \|A\|\|B\|$, and we will see later this makes $\mathscr{B}(\mathscr{H})$ into a *Banach algebra*, as will be formally defined in Chapter 5.

We also have an *involution* * of $\mathscr{B}(\mathscr{H})$; this is a map $A \mapsto A^*$ of $\mathscr{B}(\mathscr{H})$ into itself satisfying $(A^*)^* = A$, $(AB)^* = B^*A^*$, $(\alpha A + B)^* = \overline{\alpha}A^* + B^*$ for all A, B in $\mathscr{B}(\mathscr{H})$ and scalars α. Of course, * is just our adjoint operation. It is connected to the norm by $\|A^*A\| = \|A\|^2$, the result of Proposition 2.14. As we will see in Chapter 5, all of these properties together say that $\mathscr{B}(\mathscr{H})$ is a C^*-*algebra*. The relationship $\|A^*A\| = \|A\|^2$ is called the C^*-*identity*. Because there is a multiplicative identity (the identity operator I in $\mathscr{B}(\mathscr{H})$), we will say that $\mathscr{B}(\mathscr{H})$ is a *unital* C^*-algebra. Later we will make a study of C^*-algebras in general, and $\mathscr{B}(\mathscr{H})$ will be one of our primary examples. Since $\mathscr{B}(\mathscr{H})$ is noncommutative (except in the trivial situation that the dimension of $\mathscr{H}$ is 1), we will find it convenient to single out for special scrutiny the operators in $\mathscr{B}(\mathscr{H})$ which do commute with their adjoints.

Definition 2.15. An operator A in $\mathscr{B}(\mathscr{H})$ is *normal* if $AA^* = A^*A$, and *self-adjoint* if $A = A^*$.

Any multiplication operator M_φ on $L^2(X, \mu)$ is normal, since $M_\varphi^* = M_{\overline{\varphi}}$. In Chapter 6, we will see multiplication operators are, in a natural sense, the canonical examples of normal operators.

As noted above, we will be primarily interested in adjoints of operators in $\mathscr{B}(\mathscr{H})$, as opposed to $\mathscr{B}(\mathscr{H},\mathscr{K})$. One exception is if $\mathscr{H}$ and $\mathscr{K}$ are Hilbert spaces and $U : \mathscr{H} \to \mathscr{K}$ is a linear surjection that preserves inner products, meaning $\langle Uh_1, Uh_2 \rangle = \langle h_1, h_2 \rangle$ for all h_1 and h_2 in $\mathscr{H}$. Such a map is called a *Hilbert space isomorphism*.

Proposition 2.16. *If $U : \mathscr{H} \to \mathscr{K}$ is an isomorphism, then $U^*U = I_{\mathscr{H}}$ (the identity on $\mathscr{H}$) and $UU^* = I_{\mathscr{K}}$.*

Proof. Let h and g be in $\mathscr{H}$. We have

$$\langle U^*Uh, g \rangle = \langle Uh, Ug \rangle = \langle h, g \rangle$$

so that $\langle U^*Uh - h, g \rangle = 0$. This says that for a fixed h, $U^*Uh - h$ is orthogonal to every vector in $\mathscr{H}$, and hence $U^*U = I_{\mathscr{H}}$.

Now let k be in $\mathscr{K}$. Since U is surjective we may find h with $Uh = k$. Thus $UU^*k = (UU^*)Uh = Uh = k$, which gives the desired statement about UU^*. $\square$

Definition 2.17. An operator A in $\mathscr{B}(\mathscr{H},\mathscr{K})$ is said to be *invertible* if there exists B in $\mathscr{B}(\mathscr{K},\mathscr{H})$ with $AB = I_{\mathscr{K}}$ and $BA = I_{\mathscr{H}}$. We write $B = A^{-1}$.

We can rephrase the last proposition as "If U is an isomorphism, then $U^* = U^{-1}$." An isomorphism between Hilbert spaces is called a unitary operator, although some authors will restrict this terminology to the setting $\mathscr{H} = \mathscr{K}$, and some, as is our practice, will use it more generally for an isomorphism of one Hilbert space $\mathscr{H}$ onto another, possibly different, Hilbert space $\mathscr{K}$.

Definition 2.18. If $\mathscr{H}$ and $\mathscr{K}$ are Hilbert spaces and if $U : \mathscr{H} \to \mathscr{K}$ is a bijective linear map with

$$\langle Uh_1, Uh_2 \rangle_{\mathscr{K}} = \langle h_1, h_2 \rangle_{\mathscr{H}}$$

for all h_1 and h_2 in $\mathscr{H}$, then U is said to be a *unitary operator*.

It is an easy consequence of the polarization identity that a linear and surjective isometry from $\mathscr{H}$ to $\mathscr{K}$ is unitary; see Exercise 2.9.

Let us make a few elementary observations about invertibility of operators. First note that when $A \in \mathscr{B}(\mathscr{H},\mathscr{K})$ is invertible with inverse B, A must be one-to-one, since if $Ah_1 = Ah_2$ we must have $h_1 = BAh_1 = BAh_2 = h_2$. The operator A must also be onto $\mathscr{K}$, since for any $k \in \mathscr{K}$, $A(Bk) = k$. We'll discuss a converse to these statements a bit later. By linearity, A is one-to-one if and only if its *kernel*

$$\ker A \equiv \{h \in \mathscr{H} : Ah = 0\}$$

consists of the zero vector only.

We can add another property to our list of properties of $*$:

Proposition 2.19. *If A is invertible, then so is A^*, and $(A^*)^{-1} = (A^{-1})^*$.*

Proof. When A is invertible we have $AA^{-1} = I = A^{-1}A$. Applying the $*$ operation we have $(AA^{-1})^* = I^* = (A^{-1}A)^*$; clearly $I^* = I$ so that by property (d) of Proposition 2.13 we have $(A^{-1})^*A^* = I = A^*(A^{-1})^*$, which is the desired conclusion. $\square$

The next result is the converse to Proposition 2.16.

Proposition 2.20. *If U is in $\mathscr{B}(\mathscr{H}, \mathscr{K})$ with U invertible and $U^{-1} = U^*$, then U is an isomorphism.*

Proof. We have already observed that U must be surjective, so we only need to check that it preserves inner products:

$$\langle Uh, Ug \rangle = \langle h, U^*Ug \rangle = \langle h, U^{-1}Ug \rangle = \langle h, g \rangle$$

for all h and g in $\mathscr{H}$. $\square$

A few examples are in order.

Example 2.21. Let S be the forward shift on ℓ^2, so that S^* is the backward shift. We have $S^*S = I$, but S is not unitary, since its not surjective. This example points out an important distinction with the finite-dimensional situation. For a linear map T from $\mathbb{C}^n$ into itself, T is necessarily bijective if it is either one-to-one or surjective.

Example 2.22. Consider a multiplication operator M_φ on $L^2(X, \mu)$ for φ in L^∞ (X, μ). When is M_φ unitary? We want $M_\varphi M_\varphi^{-1} = M_\varphi M_\varphi^* = I$. We know that $M_\varphi^* = M_{\bar{\varphi}}$, so that M_φ is unitary if and only if $|\varphi|^2 f = f$ for all f in $L^2(X, \mu)$; that is, if and only if $|\varphi| = 1$ μ-almost everywhere.

Example 2.23. Let F map $L^2([0, 2\pi], dt/(2\pi))$ into

$$\ell^2(\mathbb{Z}) \equiv \{\{a_n\}_{-\infty}^\infty : \sum_{n=-\infty}^\infty |a_n|^2 < \infty\}$$

by $F(f) = \{\hat{f}(n)\}_{-\infty}^\infty$ where

$$\hat{f}(n) = \langle f, e^{int} \rangle = \int_0^{2\pi} f(t)e^{-int} \frac{dt}{2\pi}.$$

Linearity of F follows from linearity of the integral. Since $\{e^{int}\}$ is an orthonormal basis for $L^2([0, 2\pi], dt/(2\pi))$, part (f) of Theorem 1.33 guarantees that F preserves inner products. Given a sequence $\{a_n\}_{-\infty}^\infty \in \ell^2(\mathbb{Z})$, define $f \equiv \sum_{-\infty}^\infty a_n e^{int}$. This sum converges in $L^2([0, 2\pi])$ and $F(f) = \{a_n\}_{-\infty}^\infty$. Thus F is surjective, and hence is a unitary map.

Suppose $A \in \mathscr{B}(\mathscr{H}, \mathscr{K})$, and think of A as a mapping from $\mathscr{H}$ to $\mathscr{K}$ in the purely set-theoretic sense. From this point of view, A is invertible (as a mapping between sets) if and only if A is bijective. Its not hard to show (see Exercise 2.7) that if A is bijective, linearity of A implies that this set-theoretic inverse is a *linear* map from $\mathcal{K}$ onto $\mathcal{H}$. However, it is not at all clear that this linear set-theoretic inverse should be *bounded*. Remarkably, it is, as we will see in Section 3.3, and we are left with the extremely useful conclusion that a bounded linear operator in $\mathcal{B}(\mathcal{H},\mathcal{K})$ is invertible (in the sense of Definition 2.17) if and only if it is bijective.

We explore this a bit further. Our immediate goal is to show that we can characterize the invertible operators by weakening the requirement that A be onto if we simultaneously strengthen the requirement that A be one-to-one. The next definition describes this strengthening.

Definition 2.24. We say that $A \in \mathcal{B}(\mathcal{H},\mathcal{K})$ is *bounded below* if there is a $\delta > 0$ such that $\|Ah\| \geq \delta\|h\|$ for all h in $\mathcal{H}$.

Clearly, if A is bounded below, its kernel is $\{0\}$ and hence A is one-to-one. However, A being one-to-one does not imply that A is bounded below; a diagonal operator with diagonal sequence $\{1/n\}$ provides a counterexample.

A weakening of the condition "A maps $\mathcal{H}$ onto $\mathcal{K}$" is the requirement that the range of A is dense in $\mathcal{K}$; i.e., the closure of the range of A should be all of $\mathcal{K}$.

Theorem 2.25. *If A is a bounded linear operator from a Hilbert space $\mathcal{H}$ to a Hilbert space $\mathcal{K}$, then A is invertible if and only if A is bounded below and has dense range.*

Proof. The "only if" direction is easy: A invertible guarantees that the range of A is equal to $\mathcal{K}$, and moreover for any $h \in \mathcal{H}$,

$$\|h\| = \|A^{-1}Ah\| \leq \|A^{-1}\|\|Ah\|$$

so that

$$\|Ah\| \geq \frac{1}{\|A^{-1}\|}\|h\|$$

and A is bounded below. The "if" direction is outlined in Exercise 2.12. □

When we ask why a particular operator fails to be invertible, it is sometimes more useful to see which of the properties "bounded below" and/or "dense range" it fails to have, rather than looking at the properties "one-to-one" and "onto."

2.3 Adjoints of Banach Space Operators

So far we have defined A^* when A is a bounded linear operator between Hilbert spaces, and our definition seems closely tied to the inner product structure. We pause briefly in this section to see if we can define the adjoint of a bounded linear operator between Banach spaces. For simplicity, we restrict attention to $A \in \mathcal{B}(X)$, where X is a Banach space.

Let us begin by rephrasing the defining property of the adjoint of a Hilbert space operator. If $A \in \mathcal{B}(\mathcal{H})$ where $\mathcal{H}$ is a Hilbert space, then A^* is the unique bounded linear operator on $\mathcal{H}$ satisfying

$$\langle Ax, y \rangle = \langle x, A^* y \rangle \tag{2.3}$$

for all x and y in $\mathscr{H}$. Let the linear functional $x \to \langle x, y \rangle$ on $\mathscr{H}$ (for fixed y) be denoted by Λ_y. Thus we can rewrite Equation (2.3) as

$$\Lambda_y(Ax) = \Lambda_{A^*y}(x).$$

Even more suggestively, let us think of the map that associates the linear functional Λ_y to the linear functional Λ_{A^*y}. We have

$$\Lambda_{A^*y}(x) = \langle x, A^* y \rangle = \langle Ax, y \rangle = \Lambda_y(Ax) = (\Lambda_y \circ A)(x)$$

for x and y in $\mathscr{H}$. This suggests we try defining A^* for $A \in \mathscr{B}(X)$, where X is now a Banach space, as the map on the dual space X^* that sends $\Lambda \in X^*$ to $\Lambda \circ A$:

$$A^*(\Lambda) = \Lambda \circ A.$$

It is easy to see that $A^*(\Lambda)$ is in X^* and that the map $A^* : X^* \to X^*$ is linear.

If we adopt this as the definition of A^* when A is a bounded linear operator on the Banach space X, how does it compare with our earlier definition in the case that X is actually a Hilbert space $\mathscr{H}$? To answer this, let $C : \mathscr{H} \to \mathscr{H}^*$ be the surjective conjugate linear isometry sending y to Λ_y; "conjugate linear" referring to the fact that $C(\alpha y) = \overline{\alpha} C(y)$ for scalars α. We claim that $CA^*_{HS} = A^*_{BS}C$, as schematically illustrated below. Here the subscripts HS and BS indicate we are using the adjoint definition in, respectively, the Hilbert space setting or Banach space setting, so that A^*_{HS} acts on $\mathscr{H}$ while A^*_{BS} acts on $\mathscr{H}^*$.

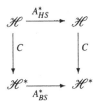

This is verified by observing that $CA^*_{HS}x$ is the bounded linear functional on $\mathscr{H}$ given as inner product with $A^*_{HS}x$, that is, the bounded linear functional taking $y \in \mathscr{H}$ to $\langle y, A^*_{HS}x \rangle = \langle Ay, x \rangle$. On the other hand, $A^*_{BS}Cx$ is the bounded linear functional on $\mathscr{H}$ taking y to

$$[A^*_{BS}(Cx)](y) = Cx(Ay) = \langle Ay, x \rangle.$$

The conjugate linearity of C means that while $(\alpha A_{HS})^* = \overline{\alpha} A^*_{HS}$, $(\alpha A)^*_{BS} = \alpha A^*_{BS}$.

It is pleasant to import the Hilbert space notation $\langle \cdot, \cdot \rangle$ into the Banach space setting. For X a Banach space, denote a generic element of X^* by x^* and write

$$x^*(x) \equiv \langle x, x^* \rangle.$$

Notice, then, that if A is in $\mathscr{B}(X)$ and A^* is in $\mathscr{B}(X^*)$, we have $\langle Ax, x^* \rangle = \langle x, A^* x^* \rangle$ since both are equal to $x^*(Ax)$.

Example 2.26. Consider the Banach space $X = C[0,1]$ in the supremum norm. As in Exercise 2.3, let φ be a continuous map of $[0,1]$ into $[0,1]$ and define the bounded linear operator C_φ on X by $C_\varphi(f) = f \circ \varphi$. Fix a point $p \in [0,1]$ and let Λ_p be the bounded linear functional of evaluation at p: $\Lambda_p(f) = f(p)$ for all $f \in X$. We seek to identify $C_\varphi^*(\Lambda_p)$ as an element of X^*. We have

$$C_\varphi^*(\Lambda_p)(f) = \Lambda_p(C_\varphi(f)) = \Lambda_p(f \circ \varphi) = f(\varphi(p)) = \Lambda_{\varphi(p)}(f)$$

for every f in X. Thus $C_\varphi^*(\Lambda_p) = \Lambda_{\varphi(p)}$. In our "inner product" notation the relevant calculation looks like

$$\langle f, C_\varphi^*(\Lambda_p) \rangle = \langle f \circ \varphi, \Lambda_p \rangle = f(\varphi(p)) = \langle f, \Lambda_{\varphi(p)} \rangle.$$

The concept of the adjoint operator had its beginnings in the work of Riesz in 1909 for operators on $L^p[a,b]$. Riesz used the terminology "Transponierte" or "transposed operator." By 1930 the idea had been extended by Banach and Juliusz Schauder to the general setting of a bounded linear operator between Banach spaces, with the terms "opération adjointe" and "opération conjuguée" being introduced.

2.4 Exercises

2.1. Let X and Y be normed linear spaces, and let $\mathscr{B}(X,Y)$ denote the collection of all bounded linear operators from X into Y endowed with the operator norm. Show that $\mathscr{B}(X,Y)$ is a normed linear space, and $\mathscr{B}(X,Y)$ is a Banach space whenever Y is a Banach space. The vector operations in $\mathscr{B}(X,Y)$ are to be defined pointwise: $(A+B)(x) = Ax + Bx$, and $(\alpha A)(x) = \alpha(Ax)$.

2.2. Suppose M is a dense subspace in a Banach space X (meaning that the closure of M is all of X) and suppose that $T : M \to Y$ is linear, where Y is a Banach space, with $\|Tm\|_Y \leq K\|m\|_X$ for some $K < \infty$ and all $m \in M$. Show that T extends, in a unique way, to a bounded linear operator from X into Y.

2.3. Let $X = [0,1]$ or more generally any compact Hausdorff space, and let $\mathscr{Y} = C(X)$, the Banach space of continuous, complex-valued functions on X, in the supremum norm. For any continuous function φ mapping X into itself, define the *composition operator* C_φ on $\mathscr{Y}$ by $C_\varphi(f) = f \circ \varphi$. Prove that C_φ is a bounded linear operator on $\mathscr{Y}$. For which φ is C_φ invertible?

2.4. Compute the norm of the multiplication operator M_z (equivalently the Toeplitz operator T_z) on $L_a^2(\mathbb{D})$.

2.5. Show that the Toeplitz operator with symbol φ acting on the Bergman space $L_a^2(\mathbb{D})$ has adjoint $T_{\overline{\varphi}}$.

2.6. Suppose T is a bounded linear operator on a Hilbert space $\mathscr{H}$ and suppose further that the range of T is one-dimensional. Show that there are vectors x and y in $\mathscr{H}$ so that

$$Tz = \langle z, x \rangle y$$

for all $z \in \mathscr{H}$. This operator is sometimes written as $y \otimes x$. Identify T^* in this case.

2.7. Show that if $T : X \to Y$ is a bijective linear map, then the set-theoretic inverse T^{-1} is also linear.

2.8. (a) Suppose that h is a nonzero vector in a Hilbert space $\mathscr{H}$. Show that $T \in \mathscr{B}(\mathscr{H})$ attains its norm at h (meaning $\|Th\| = \|T\|\|h\|$) if and only if $T^*Th = \|T\|^2 h$.
(b) Extend (a) to the case that T is a bounded linear operator from a Hilbert space $\mathscr{H}$ to a Hilbert space $\mathscr{K}$, and then use this result to provide another proof of Proposition 2.16.

2.9. Show that a linear surjective isometry from one Hilbert space $\mathscr{H}$ to another Hilbert space $\mathscr{K}$ is unitary.

2.10. Let $\{a_n\}_1^\infty$ be a bounded sequence of complex numbers. Fix an orthonormal basis $\{g_n\}$ for ℓ^2. The unique linear operator W satisfying $W(g_n) = \alpha_n g_{n+1}$ for all n is called a *weighted shift*.

(a) Find $\|W\|$ and W^*.
(b) Suppose $\{a_n\}$ and $\{b_n\}$ are bounded sequences with $|a_n| = |b_n|$ for all n. Let W and V be the associated weighted shifts. Show that there is a unitary $U : \ell^2 \to \ell^2$ with $U^{-1}WU = V$. We say that W and V are unitarily equivalent.

2.11. (a) Show that if $\sum_{n=0}^\infty |a_n|^2 (n!)^2 < \infty$, then the power series $\sum_{n=0}^\infty a_n z^n$ converges for all $z \in \mathbb{C}$ and hence $f(z) \equiv \sum_{n=0}^\infty a_n z^n$ is analytic in $\mathbb{C}$. Define the vector space of entire functions

$$\mathscr{V} \equiv \{f = \sum_{n=0}^\infty a_n z^n : \sum_{n=0}^\infty |a_n|^2 (n!)^2 < \infty\}$$

and put an inner product on $\mathscr{V}$ by setting

$$\langle f, g \rangle = \sum_{n=0}^\infty a_n \overline{b_n} (n!)^2$$

when $f(z) = \sum_{n=0}^\infty a_n z^n$ and $g(z) = \sum_{n=0}^\infty b_n z^n$. Show that $\mathscr{V}$ is a Hilbert space.
(b) Let $U : H^2 \to \mathscr{V}$ by

$$Uf = \sum_{n=0}^\infty \frac{a_n}{n!} z^n$$

when $f(z) = \sum_{n=0}^\infty a_n z^n$ is in H^2 (so that $\sum_{n=0}^\infty |a_n|^2 < \infty$; see Example 1.7). Recalling that the power series coefficients a_n of f are given by

$$a_n = \frac{f^{(n)}(0)}{n!}$$

show that U is a unitary map.

(c) Define a linear map D on $\mathscr{V}$ by $Df = f'$, and show that D is a bounded linear operator on $\mathscr{V}$.

(d) Let $B : H^2 \to H^2$ be defined by

$$Bf = \frac{f - f(0)}{z}.$$

Show that B is a bounded linear operator on H^2 and $D = UBU^{-1}$, so that D and B are unitarily equivalent.

2.12. Suppose that $A \in \mathscr{B}(\mathscr{H}, \mathscr{K})$ where A is bounded below and has dense range in $\mathscr{K}$. Show that A is invertible. Hint: Start by showing that if A is bounded below, then the range of A is closed.

2.13. Suppose $A_n \in \mathscr{B}(\mathscr{H}_n)$ for $n = 1, 2, 3, \ldots$, where each $\mathscr{H}_n$ is a Hilbert space. Assume further that $\sup_n \|A_n\| < \infty$. Define A on $\mathscr{H} \equiv \sum \oplus \mathscr{H}_n$ by

$$A(h) = A(h_1, h_2, \ldots) = (Ah_1, Ah_2, \ldots).$$

Show that $A \in \mathscr{B}(\mathscr{H})$ and $\|A\| = \sup_n \|A_n\|$. We call A the direct sum of the operators $\{A_n\}$ and denote it $\sum \oplus A_n$.

2.14. Suppose that u is a bounded sesquilinear form on $\mathscr{H} \times \mathscr{H}$ for some Hilbert space $\mathscr{H}$.

(a) Define the *quadratic form* $\hat{u} : \mathscr{H} \to \mathbb{C}$ by $\hat{u}(h) = u(h, h)$. Show the polarization identity

$$u(h, g) = \hat{u}\left(\frac{1}{2}(h+g)\right) - \hat{u}\left(\frac{1}{2}(h-g)\right) + i\hat{u}\left(\frac{1}{2}(h+ig)\right) - i\hat{u}\left(\frac{1}{2}(h-ig)\right)$$

for all h and g in $\mathscr{H}$.

(b) Show that if u_1 and u_2 are bounded sesquilinear forms on $\mathscr{H} \times \mathscr{H}$ with $u_1(h, h) = u_2(h, h)$ for all h in $\mathscr{H}$ then u_1 and u_2 agree on $\mathscr{H} \times \mathscr{H}$.

2.15. Recall the space $C^1[0, 1]$ defined by

$$C^1([0, 1]) \equiv \{f \in C[0, 1] : f' \text{ exists and is continuous on } [0, 1]\}.$$

(a) Recall from Exercise 1.3 of Chapter 1 that $C^1[0, 1]$ is a Banach space if we norm it by

$$\|f\|_{C^1} = \max\{|f(x)| : 0 \leq x \leq 1\} + \max\{|f'(x)| : 0 \leq x \leq 1\}.$$

Show that for each $x \in [0, 1]$, the functional

$$ev_x^1(f) = f'(x)$$

is a bounded linear functional on $(C^1[0,1], \|\cdot\|_{C^1})$.

(b) Fix $\varphi : [0,1] \to [0,1]$ in $C^1[0,1]$ and define C_φ on $C^1[0,1]$ by $C_\varphi(f) = f \circ \varphi$. Show that C_φ is a bounded linear operator on $C^1[0,1]$. What is $C_\varphi^*(ev_x^1)$?

2.16. For an operator T in $\mathscr{B}(\mathscr{H})$, where $\mathscr{H}$ is a Hilbert space, show that $\ker T = (\operatorname{ran} T^*)^\perp$, where $\ker T = \{h : Th = 0\}$ and $\operatorname{ran} T^* \equiv \operatorname{range} T^* = \{T^*h : h \in \mathscr{H}\}$.

2.17. If T is a bounded and self-adjoint operator on a Hilbert space and $T^2 = T$, show that T is the orthogonal projection onto its range.

2.18. Suppose that P_{M_1} and P_{M_2} are orthogonal projections onto the closed subspaces M_1 and M_2 of a Hilbert space.

(a) Show that $P_{M_1}P_{M_2}$ is an orthogonal projection if and only if $P_{M_1}P_{M_2} = P_{M_2}P_{M_1}$.
(b) If the condition in (a) is satisfied so that $P_{M_1}P_{M_2} = P_M$ for some closed subspace M, identify M.

2.19. Show that a diagonal operator on a Hilbert space is an orthogonal projection if and only if its diagonal consists of 0's and 1's.

2.20. Fix vectors $h_1, h_2, \ldots, h_n$ in a Hilbert space $\mathscr{H}$.

(a) Define $B : \mathbb{C}^n \to \mathscr{H}$ by

$$B(z_1, z_2, \ldots, z_n) = \sum_{j=1}^n z_j h_j.$$

Calculate $B^* : \mathscr{H} \to \mathbb{C}^n$.

(b) What is the relationship between the operator B in (a), the $n \times n$ matrix $A = [a_{ij}]$ with $a_{ij} = \langle h_j, h_i \rangle$, and the operator $C : \mathscr{H} \to \mathscr{H}$ given by

$$Ch = \sum_{j=1}^n \langle h, h_j \rangle h_j?$$

(c) Show that $h_1, h_2, \ldots, h_n$ are linearly independent in $\mathscr{H}$ if and only if the matrix A is invertible.

2.21. Suppose that T is an operator in $\mathscr{B}(\mathscr{H})$ for some Hilbert space $\mathscr{H}$ and suppose that $T = T^{-1}$ and $T = T^*$. Show that the sets

$$\mathscr{H}_1 \equiv \{h + Th : h \in \mathscr{H}\}$$

and

$$\mathscr{H}_2 \equiv \{h - Th : h \in \mathscr{H}\}$$

are closed subspaces of $\mathscr{H}$ with $\mathscr{H} = \mathscr{H}_1 \oplus \mathscr{H}_2$, and that the restriction of T to $\mathscr{H}_1$ is the identity I while the restriction of T to $\mathscr{H}_2$ is $-I$.

Conversely, show that if $\mathscr{H}$ is the direct sum of two subspaces $\mathscr{H}_1$ and $\mathscr{H}_2$ with $T(h) = h$ for $h \in \mathscr{H}_1$ and $T(h) = -h$ for $h \in \mathscr{H}_2$, then $T = T^{-1}$ and $T^* = T$.

2.22. If $A \in \mathcal{B}(\mathcal{H})$ and $h \in \mathcal{H}$, then

$$\{h, Ah, A^2h, A^3h, \ldots\}$$

is called the *orbit* of h under A. When the orbit of h spans $\mathcal{H}$— that is, when the set of linear combinations of vectors in the orbit of h is dense in $\mathcal{H}$—we call h a *cyclic vector* for A, and say that A is a *cyclic operator*.

(a) Show that the constant function 1 is a cyclic vector for M_z on $L_a^2(\mathbb{D})$.
(b) Consider the operator from $\mathbb{C}^3$ to $\mathbb{C}^3$ with matrix

$$A = \begin{bmatrix} 2 & 0 & 0 \\ 0 & 2 & 0 \\ 0 & 0 & 1 \end{bmatrix}.$$

Show that this operator has no cyclic vector.

Cyclic operators will be studied further in Chapter 6.

2.23. Thus far, the only topology we have considered on $\mathcal{B}(\mathcal{H})$ for $\mathcal{H}$ a Hilbert space is the topology that comes from the operator norm; this is called the *norm topology* or the *uniform operator topology*. However, there are other useful topologies on $\mathcal{B}(\mathcal{H})$, and in this problem we introduce two of them by discussing sequential convergence in two new senses.
Definition. Given $\{T_n\}$ in $\mathcal{B}(\mathcal{H})$, we say $T_n \to T \in \mathcal{B}(\mathcal{H})$ in the *strong operator topology* if

$$T_n h \to Th$$

for each $h \in \mathcal{H}$. This is abbreviated $T_n \to T$ (SOT).
    We say T_n converges to T in the *weak operator topology*, denoted $T_n \to T$ (WOT), if

$$\langle T_n h, g \rangle \to \langle Th, g \rangle$$

for each fixed h, g in $\mathcal{H}$.

(a) Show that if S is the forward shift operator on ℓ^2, then $S^n \to 0$ (WOT), but S^n does not converge to 0 in either the norm or strong operator topology.
(b) If $P_n : \ell^2 \to \ell^2$ by

$$P_n(x_1, x_2, \ldots) = (0, 0, \ldots, x_{n+1}, x_{n+2}, \ldots)$$

show that $P_n \to 0$ (SOT), but not in the norm topology.
(c) Show that if $T_n \to T$ (SOT) for $T_n, T \in \mathcal{B}(\mathcal{H})$, then $T_n \to T$ (WOT).
(d) Is the mapping from $\mathcal{B}(\mathcal{H}) \to \mathbb{C}$ which sends T to $\|T\|$ continuous if we use the strong operator topology (respectively, the weak operator topology) on $\mathcal{B}(\mathcal{H})$?

Chapter 3
The Big Three

> *In linear spaces with a suitable topology, one encounters three far-reaching principles concerning continuous linear transformations.*
> *N. Dunford and J. Schwartz ([12], p. 49).*

In this chapter we will look at several core results of functional analysis. Two of these, the principle of uniform boundedness and the open mapping theorem (as well as their close cousins, the Banach–Steinhaus theorem and the closed graph theorem) are Banach space results. The third, the Hahn–Banach theorem, makes no use of completeness and takes place in a normed linear space (or even more generally). All three of these results are ubiquitous in functional analysis.

We will look at the Hahn–Banach theorem first, and begin by reviewing Zorn's lemma, which is sometimes called the "analysts' version of the axiom of choice." We begin with some needed terminology. A *partial order* on a set X is a relation, written generically as $\leq$, satisfying the following properties for all $a, b, c \in X$:

(a) transitivity: if $a \leq b$ and $b \leq c$ then $a \leq c$,
(b) reflexivity: $a \leq a$, and
(c) anti-symmetry: if $a \leq b$ and $b \leq a$ then $a = b$.

If for every pair a, b in X we either have $a \leq b$ or $b \leq a$, then X is said to be totally ordered by $\leq$. Two simple examples to illustrate these concepts are the totally ordered set consisting of the real line with the usual $\leq$ relationship, and, for any set X, any collection of subsets of X partially ordered by set inclusion $\subseteq$. Note that the latter need not be a total ordering, for example when the cardinality of X is at least two, and $\mathscr{P}(X)$ is the set of all subsets of X, then $(\mathscr{P}(X), \subseteq)$ is a partial ordering which is not a total ordering.

As further examples, consider the set $\mathbb{N} \times \mathbb{N}$ of ordered pairs of positive integers and two relations $\leq_1$ and $\leq_2$ defined as follows. Say that $(a, b) \leq_1 (x, y)$ if a is strictly less than x or if $a = x$ and b is less than or equal to y (in the usual sense on $\mathbb{N}$). This is sometimes called lexicographical ordering, by its analogy with the ordering of words in a dictionary. For the second relation, say $(a, b) \leq_2 (x, y)$ if a is less than or equal to x, and b is less than or equal to y, in the usual sense of inequality on $\mathbb{N}$. Clearly $\leq_1$ is a total ordering, while $\leq_2$ is a partial ordering which is not a total ordering.

B.D. MacCluer, *Elementary Functional Analysis*, DOI 10.1007/978-0-387-85529-5_3,
© Springer Science+Business Media, LLC 2009

When a set X is partially ordered by $\leq$ and Y is a subset of X, we call an element $p \in X$ an *upper bound* for Y if $y \leq p$ for all $y \in Y$. An element m in X for which $m \leq x$ implies $m = x$ is called a *maximal* element of X.

We are ready to state Zorn's lemma, which is indispensable in functional analysis. It will be taken as an axiom of set theory; alternatively one can derive it from the axiom of choice, which says that the Cartesian product of nonempty sets is nonempty.

Lemma 3.1 (Zorn's Lemma). *If X is a nonempty partially ordered set with the property that every totally ordered subset of X has an upper bound in X, then X has a maximal element.*

A good way to see a standard and completely straightforward application of Zorn's lemma is to use it to prove that any orthonormal set in a Hilbert space can be extended to an orthonormal basis, so in particular every Hilbert space has an orthonormal basis. See Exercise 3.1. The proof of the Hahn–Banach theorem in the next section will provide another example of an application of Zorn's lemma.

3.1 The Hahn–Banach Theorem

The Hahn–Banach theorem deals with extending continuous linear functionals from a subspace of a normed linear space to the whole space. Completeness of the space plays no role, so this is a result about normed linear spaces in general. This is one of the few places in the subject where it is helpful to first look at *real* vector spaces (that is, a vector space over $\mathbb{R}$), and prove it in that context, before extending the argument to cover the case of a complex vector space. This extension, while not difficult, is not trivial either; historically the real case was proved nearly ten years before its extension to complex scalars.

Theorem 3.2 (Hahn–Banach Theorem). *Let X be a normed linear space over $\mathbb{F} = \mathbb{R}$ or $\mathbb{C}$, and suppose Y is a (not necessarily closed) proper subspace of X. If $\varphi_0 : Y \to \mathbb{F}$ is a bounded linear functional, then there is a bounded linear functional $\varphi : X \to \mathbb{F}$ with the restriction of φ to Y equal to φ_0 (so that φ is an extension of φ_0) and $\|\varphi\| = \|\varphi_0\|$ (so that this extension is norm-preserving).*

It is the norm-preserving part of the conclusion that gives the result its power. Recall that Exercise 1.22 in Chapter 1 showed how to extend linear functionals on subspaces of Hilbert spaces, using simple Hilbert space techniques, in a norm-preserving way. Without Hilbert space machinery at our disposal, we will have to work a bit harder.

Before we look at the proof of the Hahn–Banach theorem, we give some applications. For the first, suppose X is a normed linear space which is not just $\{0\}$. Could X^*, the dual space of X, consist of just the zero functional?

Corollary 3.3. *Let $X \neq \{0\}$ be a normed linear space. Given $x_0 \neq 0$ in X, there is a bounded linear functional φ on X of norm 1 with $\varphi(x_0) = \|x_0\|$.*

Proof. Set $M = \{\alpha x_0 : \alpha \in \mathbb{F}\}$, a subspace of X. Define φ on M by $\varphi(\alpha x_0) = \alpha\|x_0\|$. It is easy to see that φ is a bounded linear functional on M with norm 1 and $\varphi(x_0) = \|x_0\|$. By the Hahn–Banach theorem we can extend φ to all of X without increasing its norm. $\qquad\square$

The next corollary shows that the dual of a nontrivial normed linear space must be rich enough to "separate the points" of X.

Corollary 3.4. *Suppose $X \neq \{0\}$ is a normed linear space. Given $x_1 \neq x_2$ we may find a bounded linear functional φ on X with $\varphi(x_1) \neq \varphi(x_2)$.*

Proof. Apply Corollary 3.3 to $x_0 = x_1 - x_2$. $\qquad\square$

Corollary 3.5. *Suppose x_0 is an element of a normed linear space X. We have*

$$\|x_0\| = \sup\{|\varphi(x_0)| : \varphi \in X^*, \|\varphi\| = 1\},$$

and moreover this supremum is attained.

Proof. The result if trivially true if $x_0 = 0$. In general, we have $|\varphi(x_0)| \leq \|x_0\|$ if $\|\varphi\| = 1$, so the supremum is at most $\|x_0\|$. On the other hand, by Corollary 3.3 there exists $\hat{\varphi} \in X^*$ with $\|\hat{\varphi}\| = 1$ and $\hat{\varphi}(x_0) = \|x_0\|$, so the supremum must in fact be equal to $\|x_0\|$, *and* the supremum is attained (by $\hat{\varphi}$). $\qquad\square$

See Exercise 3.5 for a concrete application of Corollary 3.5 in the particular normed linear space $L^p(X, \mathfrak{M}, \mu)$, $1 \leq p < \infty$.

Note the symmetry in the statements: For $\varphi \in X^*$,

$$\|\varphi\| = \sup\{|\varphi(x)| : x \in X, \|x\| = 1\},$$

and for $x \in X$,

$$\|x\| = \sup\{|\varphi(x)| : \varphi \in X^*, \|\varphi\| = 1\}.$$

In the second line (but not in general in the first, although see Corollary 3.7 below), "sup" can be replaced by "max."

The Hahn–Banach theorem gives a means for determining the points that lie in the closure of a linear subspace of a normed linear space. The proof of the next result is left to the reader as Exercise 3.8.

Corollary 3.6. *Suppose that X is a normed linear space, $x_0 \in X$ and M is a (not necessarily closed) subspace in X. Suppose that $d \equiv \mathrm{dist}(x_0, M) > 0$ where $\mathrm{dist}(x_0, M) = \inf\{\|x_0 - y\| : y \in M\}$. There exists $\varphi \in X^*$ with $\varphi(x_0) = 1$, $\varphi = 0$ on M, and $\|\varphi\| = 1/d$. In particular, x_0 is in the closure of M if and only if there is no bounded linear functional on X that is 0 on M and nonzero at x_0.*

When X is a Banach space, or even just a normed linear space, we know from Exercise 2.1 in Chapter 2 that its dual X^* is a Banach space, so that it too has a dual space $(X^*)^*$ which we will write X^{**}. If $x_0 \in X$ we can define what will constitute an element of X^{**}, call it x_0^{**}, by setting $x_0^{**}(\varphi) = \varphi(x_0)$, for any φ in X^*. It is easy to see that x_0^{**} as just defined is a linear functional on X^* and moreover,

$$|x_0^{**}(\varphi)| = |\varphi(x_0)| \leq \|\varphi\| \|x_0\|$$

so that x_0^{**} is bounded with norm at most $\|x_0\|$. We claim that this natural map from X to X^{**} sending x_0 to x_0^{**} is a linear isometry of X into X^{**}. The interesting piece that still needs verification is the "isometry" part of this statement. We have

$$\|x_0^{**}\| = \sup\{|x_0^{**}(\varphi)| : \varphi \in X^*, \|\varphi\| = 1\}$$
$$= \sup\{|\varphi(x_0)| : \varphi \in X^*, \|\varphi\| = 1\}$$
$$= \|x_0\|$$

where we have used the definition of the norm on X^{**}, the definition of x_0^{**}, and Corollary 3.5 for each of the three equalities, respectively. If this natural map $x_0 \rightarrow x_0^{**}$ is onto X^{**}, then X is said to be reflexive, and it gives a isometric isomorphism of X with X^{**}. In this context, "isomorphism" means a continuous linear bijection with continuous inverse. If X is reflexive, it must be a Banach space since X^{**} is a Banach space. Note that the definition of "reflexive" requires that a *particular* mapping of X into X^{**} (the "natural map") be an isometric isomorphism, not simply that there exist some isometric isomorphism of X and X^{**}. The latter surprisingly turns out to be a strictly weaker assumption; this was shown by R.C. James in [23]. As an immediate consequence of the definition of a reflexive space and Corollary 3.5 we see that bounded linear functionals on reflexive Banach spaces attain their norm; this is the content of the next result, whose proof is left to the reader.

Corollary 3.7. *Suppose that X is a reflexive Banach space. Given $\varphi \in X^*$, there exists a unit vector x_0 in X such that $|\varphi(x_0)| = \|\varphi\|$.*

This corollary is sometimes useful for showing a particular Banach space is not reflexive; see, for example, Exercise 3.34. The converse to Corollary 3.7 is also true: In any nonreflexive Banach space there is a bounded linear functional which does not attain its norm on the unit sphere. This result is also due to James [24].

We are going to prove Theorem 3.2 in the real scalar case first, that is, we will assume in the statement of the theorem that $\mathbb{F} = \mathbb{R}$. There are two key parts to the proof: the "one-step extension," which basically extends a linear functional, in a norm-preserving way, from a linear manifold to the span of that manifold and a single additional vector; and a Zorn's lemma argument.

Proof (Theorem 3.2, real case). If $\|\varphi_0\| = 0$, simply set $\varphi \equiv 0$ and we are done. Thus we are interested in the case $\|\varphi_0\| \neq 0$, and we may assume, without loss of generality, that $\|\varphi_0\| = 1$, which we do. Choose a vector z which lies in X but not in Y and let

$$Y_1 \equiv \{y + \alpha z : y \in Y \text{ and } \alpha \in \mathbb{R}\} = \{\alpha z - y : y \in Y \text{ and } \alpha \in \mathbb{R}\},$$

so that Y_1 is the span of Y and z. For any choice of a fixed real number c, if we define φ_1 on Y_1 by $\varphi_1(\alpha z - y) = \alpha c - \varphi_0(y)$, then φ_1 is an extension of φ_0 to a linear functional on Y_1. There are a few details to be checked here, starting with the observation that φ_1 is well-defined (once c is chosen) and linear. Well-definedness follows from uniqueness of representation for every vector in Y_1 in the form $\alpha z - y$ with $y \in Y$ and $\alpha \in \mathbb{R}$. Once these easy issues are attended to (we leave the details to the reader), the issue becomes whether we can choose c so that $\|\varphi_1\| = \|\varphi_0\| = 1$. In other words, we want to choose c so that

$$|\alpha c - \varphi_0(y)| \leq \|\alpha z - y\| \tag{3.1}$$

for all y in Y and all scalars α in $\mathbb{R}$. Now we want (3.1) to hold for all $y \in Y$ and real scalars α, so it can be rewritten in an equivalent manner by dividing by $|\alpha|$ (since the $\alpha = 0$ case is trivially true). Doing this, we see that the condition (3.1) we wish to satisfy is equivalent to the condition

$$|c - \varphi_0(\alpha^{-1}y)| \leq \|z - \alpha^{-1}y\|$$

for all $y \in Y$ and real $\alpha \neq 0$, or more simply, to the condition

$$|c - \varphi_0(y)| \leq \|z - y\| \tag{3.2}$$

for all $y \in Y$. This last inequality will hold for some choice of c precisely when there is a choice of c satisfying

$$\varphi_0(y) - \|y - z\| \leq c \leq \varphi_0(y) + \|y - z\| \tag{3.3}$$

for all $y \in Y$. If we denote the left-hand side, $\varphi_0(y) - \|y - z\|$, by $A(y)$ and the right-hand side, $\varphi_0(y) + \|y - z\|$, by $B(y)$, a real c of the desired type can be found provided

$$\bigcap_{y \in Y} [A(y), B(y)]$$

is nonempty (and if it is nonempty, any c that lies in this intersection will do). We claim that

$$\bigcap_{y \in Y} [A(y), B(y)]$$

is nonempty precisely when $A(y) \leq B(v)$ for all y and v in Y. One direction of this claim is clear: If $c \in [A(y), B(y)]$ for all $y \in Y$, then $c \geq A(y)$ for all y and $c \leq B(v)$ for all v. Conversely, if $A(y) \leq B(v)$ for all choices of y and v, then $a \equiv \sup_{y \in Y} A(y) \leq \inf_{v \in Y} B(v) \equiv b$, and any c in the range $a \leq c \leq b$ will lie in $[A(y), B(y)]$ for all y. This verifies the claim, and we can create a norm-preserving extension if

$$\varphi_0(y) - \|y - z\| \leq \varphi_0(v) + \|v - z\| \tag{3.4}$$

for all y and v in Y. We have

$$\varphi_0(y) - \varphi_0(v) = \varphi_0(y-v) \le \|y-v\| \le \|y-z\| + \|z-v\|,$$

where we have used the assumption that φ_0 has norm 1. Rearranging this computation gives (3.4), and we conclude that there is a norm-preserving extension φ_1 of φ_0 from Y to Y_1.

To finish the proof, we use Zorn's lemma. Let $\mathscr{P}$ be the collection of all pairs (Y', φ') where Y' is a (not necessarily closed) subspace containing Y and $\varphi : Y' \to \mathbb{R}$ is a linear functional extending $\varphi_0 : Y \to \mathbb{R}$ with $\|\varphi'\| = \|\varphi_0\| = 1$. Partially order $\mathscr{P}$ by $(Y', \varphi') \le (Y'', \varphi'')$ if $Y' \subseteq Y''$ and the restriction of φ'' to Y' is φ'. Suppose $\mathscr{C} \equiv \{(Y_\beta, \varphi_\beta) : \beta \in \mathscr{I}\}$ is a totally ordered subset of $\mathscr{P}$. Set $N \equiv \cup_\beta Y_\beta$. Since $\mathscr{C}$ is totally ordered, N is a subspace. Define $\tilde{\varphi}$ on N by $\tilde{\varphi}(y) = \varphi_\beta(y)$ if $y \in Y_\beta$; note that $\tilde{\varphi}$ is well-defined since $\mathscr{C}$ is totally ordered, and is linear. Moreover, $(N, \tilde{\varphi})$ is in $\mathscr{P}$; in particular there is an β so that

$$|\tilde{\varphi}(y)| = |\varphi_\beta(y)| \le \|y\|$$

so $\|\tilde{\varphi}\| \le 1$. We see that $(N, \tilde{\varphi})$ is an upper bound for $\mathscr{C}$, since $(Y_\beta, \varphi_\beta) \le (N, \tilde{\varphi})$ for all β. By Zorn's lemma, $\mathscr{P}$ has a maximal element, which we denote $(X_\infty, \varphi_\infty)$. We must have $X_\infty = X$, else we could do the one-step extension process to extend to the span of X_∞ and x_0, where x_0 is in X but not in X_∞, contradicting the maximality of $(X_\infty, \varphi_\infty)$. Once we know that $X_\infty = X$, we have the desired norm-preserving extension φ_∞ of φ_0 to all of X. $\square$

In Exercise 3.4, the reader can work through an application of the one-step extension process in a concrete setting.

The extension of the proof of the Hahn–Banach theorem from the real case to the complex case is outlined in Exercises 3.2 and 3.3. While it is not hard, historically there was a span of nearly ten years between the work on the real case by Banach and the extension to the complex case by H. Bohnenblust and A. Sobczyk in 1938. Perhaps not coincidently, Banach's esteemed 1932 treatise *Opérations Linéaires* deals only with *real* Banach spaces. In the particular setting of $X = L^p$ the complex case appeared in 1936 in the work of F. Murray; see also the comment in Exercise 3.3 on an earlier contribution by H. Löwig. The work of Bohenblust and Sobczyk may be the first place that the result is referred to as the "Hahn–Banach Theorem."

In reality, it would be more accurate to credit Eduard Helly with the first proof of a Hahn–Banach type theorem for work dating from 1912. Helly was working on a problem, posed by Riesz, which he reformulated as a problem about extending a bounded linear functional on a subspace of $C[a,b]$. His argument had a thoroughly modern flavor, and was quite similar to that later used independently by Hahn (1927) and Banach (1929). A key feature was the one-step extension process, and in particular the inequalities of (3.3) and (3.4) for the special case of $C[a,b]$ appears in his work. A slightly later paper of Helly's [22], published in 1921, gives the Hahn–Banach theorem in the context of sequence spaces.

One possible explanation for the lack of recognition Helly received for this (and other mathematical contributions he made) might be found in some of the non-mathematical details of his life (see [20]). An Austrian, he received his Ph.D. in 1907. He enlisted in the army at the start of World War I, and was wounded in 1915, suffering serious heart and lung injuries. He became a Russian prisoner of war, and was imprisoned in Siberia from 1915–1917. In part due to the civil war in Russia in 1918, he was not able to return to his home of Vienna until late in 1920. He received his Habilitation degree from the University of Vienna in 1921, but was unable to secure an academic position. He worked as a bank clerk (until the bank failed in 1929) and in an insurance company, while trying to remain active in mathematics. Helly—who was Jewish—emigrated to the United States in 1938, as Austria was absorbed by the Third Reich. He held positions at several junior colleges in New Jersey until he was offered a Professorship at the Illinois Institute of Technology. Unfortunately, shortly after this he died of a heart attack, at the age of 59. Undoubtedly his earlier war injuries contributed to this premature death.

3.2 Principle of Uniform Boundedness

Several important problems in Banach space theory come down to the rather pedestrian-seeming problem of showing that a set is "large" in the sense that it has nonempty interior. Recall that the interior of a set A in a metric space (M,d) is the set of points $a \in A$ for which there exists $\delta > 0$ with $B(a,\delta) \equiv \{m \in M : d(a,m) < \delta\} \subseteq A$. As an illustration of this general principle, let us characterize boundedness of a linear operator in terms of a particular set having nonempty interior.

Proposition 3.8. *Suppose X and Y are normed linear spaces and $T : X \to Y$ is linear. Then T is bounded if and only if $T^{-1}(\{y \in Y : \|y\| < 1\})$, the preimage of the open unit ball in Y under T, has nonempty interior.*

Proof. First suppose T is bounded with $\|T\| = M$. If $x \in B(0, 1/M)$ we have $\|Tx\| < 1$, and thus $B(0, 1/M)$ is contained in the preimage of the open unit ball of Y under T.

The more interesting direction is the "if" direction. For this, suppose T is linear and $T^{-1}\{y : \|y\|_Y < 1\}$ contains x_0 as an interior point, say $B(x_0, \varepsilon)$ lies in the preimage of the open unit ball in Y under T. Fix $\varepsilon', 0 < \varepsilon' < \varepsilon$, and consider x with $\|x\| \leq \varepsilon'$. Since $x + x_0 \in B(x_0, \varepsilon)$, we have

$$\|Tx\| = \|T(x + x_0 - x_0)\| = \|T(x + x_0) - T(x_0)\|$$
$$\leq \|T(x + x_0)\| + \|Tx_0\|$$
$$\leq 1 + \|T(x_0)\| \equiv M.$$

Thus for any unit vector v in X,

$$\|Tv\| = \left\| T\left(\frac{1}{\varepsilon'} \varepsilon' v \right) \right\| \leq \frac{M}{\varepsilon'}$$

and T is bounded. □

So when does a set have nonempty interior? In a complete metric space (thus in particular in a Banach space) the Baire category theorem sheds some light on this.

Definition 3.9. A set S in a metric space M is *nowhere dense* if its closure has empty interior.

Some examples of nowhere dense sets are the integers $\mathbb{Z}$ in the real line $\mathbb{R}$, or the Cantor set in $[0,1]$ (it is a closed set containing no open intervals). By contrast, the rationals are *not* a nowhere dense set in $\mathbb{R}$.

The next result is the Baire category theorem. Its roots are in René Baire's 1899 dissertation, where it was shown that $\mathbb{R}^n$ is not a countable union of nowhere dense sets (the case $n = 1$ was actually proved two years earlier by W. Osgood). One can think of this as a topological tool; it will play a role in the proof of the principle of uniform boundedness of this section, and the proof of the open mapping theorem in the next.

Theorem 3.10 (Baire Category Theorem). *A complete metric space is not the union of a countable number of nowhere dense sets.*

Proof. Let M be a complete metric space. Suppose, for a contradiction, that M is the countable union of sets A_n that are nowhere dense. We will construct a Cauchy sequence in M with no limit point in M.

Since A_1 is nowhere dense, we may find an open ball B_1 with $B_1 \cap \overline{A_1} = \emptyset$, where $\overline{A_1}$ denotes the closure of A_1. Clearly, we can choose the radius of this ball B_1 to be less than 1. Since $\overline{A_2}$ has empty interior, it doesn't contain B_1 and $(M \backslash \overline{A_2}) \cap B_1$ is a nonempty open set. Thus we may find an open ball B_2 whose closure is contained in B_1 such that $B_2 \cap \overline{A_2} = \emptyset$ and such that the radius of B_2 is less than $1/2$. Continue inductively, so that at the nth stage we produce an open ball B_n whose closure is contained in B_{n-1}, such that $\underline{B_n} \cap \overline{A_n} = \emptyset$ and the radius of B_n is less than $1/n$; we are using the hypothesis that $\overline{A_n}$ has no interior point and thus B_{n-1} is not contained in $\overline{A_n}$. Note that the closed balls $\overline{B_n}$ form a nested sequence of closed sets, whose diameters tend to 0, in our complete metric space.

Let x_n be the center of the ball B_n. It is easy to see that $\{x_n\}$ forms a Cauchy sequence in M: When $m, n \geq N$, the points x_m and x_n lie in the ball B_N and hence $d(x_m, x_n) < 2/N$, which tends to 0 as $N \to \infty$. This Cauchy sequence then converges to some point $x \in M$. We claim that x is not in A_j for all $j \geq 1$, a contradiction to the assumption that $M = \cup_{j=1}^{\infty} A_j$. To verify this claim, note that if $x \in A_j$, then x is not in B_j (since $B_j \cap \overline{A_j} = \emptyset$), and hence not in $\overline{B_k}$ for all $k \geq j + 1$. However, we have $x_n \in B_{j+1}$ for all $n \geq j + 1$ and therefore x is in $\overline{B_{j+1}} \subseteq B_j$, a contradiction. □

Thus the Baire category theorem says that if a complete metric space X is written as a countable union $X = \cup_1^{\infty} A_n$, at least one of the sets A_n must be "big" in the sense that its closure has nonempty interior. The proof of the next result, the principle of uniform boundedness, illustrates this. Informally, this theorem says that a pointwise bounded family of operators is uniformly bounded.

Theorem 3.11 (Principle of Uniform Boundedness). *Suppose X is a Banach space and $\mathscr{F}$ is a family of bounded linear operators from X to a normed linear space Y. If, for every $x \in X$,*

$$\sup\{\|Tx\| : T \in \mathscr{F}\} < \infty$$

then

$$\sup\{\|T\| : T \in \mathscr{F}\} < \infty.$$

Proof. Define $A_n \equiv \{x \in X : \|Tx\| \leq n \text{ for all } T \in \mathscr{F}\}$. The hypothesis says that each $x \in X$ is in some A_n, so that $X = \cup_{n=1}^{\infty} A_n$. By the Baire category theorem, for some n, $\overline{A_n}$ has nonempty interior. To make use of this we first claim that each A_n is closed. To see this, suppose x_m is in A_n for $m = 1, 2, \ldots$ and that $x_m \to x$. For each $T \in \mathscr{F}$, $\|Tx_m\| \leq n$, while continuity of T guarantees that $\|Tx_m\| \to \|Tx\|$. Hence $\|Tx\| \leq n$ for each $T \in \mathscr{F}$ and $x \in A_n$.

So in fact, for some fixed n, A_n has an interior point, which we will denote x_0. Let $\varepsilon > 0$ be chosen so that $\overline{B(x_0, \varepsilon)} \subseteq A_n$. The positive number ε, as well as the integer n, are fixed values at this point. If $\|x\| \leq \varepsilon$, then for any $T \in \mathscr{F}$,

$$\|Tx\| = \|T(x+x_0) - Tx_0\| \leq \|T(x+x_0)\| + \|Tx_0\| \leq n+n.$$

From this it follows that for any unit vector v,

$$\|Tv\| = \frac{1}{\varepsilon}\|T(\varepsilon v)\| \leq \frac{2n}{\varepsilon}$$

and thus

$$\sup\{\|T\| : T \in \mathscr{F}\} \leq \frac{2n}{\varepsilon}.$$

$\qquad\qquad\square$

Note that we can restate the principle of uniform boundedness as follows: either $\sup\{\|T\| : T \in \mathscr{F}\} < \infty$, or there exists $x \in X$ such that $\sup\{\|Tx\| : T \in \mathscr{F}\} = \infty$.

As with the Hahn–Banach theorem, Helly deserves more credit than he has received for his contributions to the uniform boundedness principle. He gave the first proof, for $C[a, b]$, but by methods which extend to general Banach spaces. Banach and Steinhaus's original proof depended on a technique called the "gliding hump" method; this was replaced by the Baire category argument after S. Saks pointed out the possibility of using this approach. The original gliding hump argument is outlined in Exercise 3.15. A precursor to the principle of uniform boundedness, in the setting of ℓ^2, appeared in work of E. Hellinger and O. Toeplitz in 1910.

A close cousin of the principle of uniform boundedness is the Banach–Steinhaus theorem, which we look at next.

Theorem 3.12 (Banach–Steinhaus Theorem). *Suppose $\{T_n\}$ is a sequence of bounded linear operators from a Banach space X to a Banach space Y. Assume further that for all $x \in X$, $\lim_{n \to \infty} T_n x$ exists. Define $T : X \to Y$ by $Tx = \lim_{n \to \infty} T_n x$. With this definition, T is a bounded linear operator from X to Y.*

Proof. It is easy to check that T is linear, and we leave the details of this to the reader. We will use the principle of uniform boundedness to show that T is bounded. For each $x \in X$, $\sup_n \|T_n x\| < \infty$ since $\{T_n x\}$ is a convergent sequence by hypothesis and convergent sequences in a metric space are bounded. By Theorem 3.11, $\sup_n \|T_n\| \equiv M < \infty$. This means that for each $x \in X$, and any n, $\|T_n x\| \leq M\|x\|$. We have

$$\|Tx\| = \|\lim_{n \to \infty} T_n x\| \leq M\|x\|$$

(note the continuity of the norm lurking behind this calculation) and T is bounded with $\|T\| \leq M$. $\qquad\square$

Note that the last result does not say that if $T_n \to T$ pointwise on X, then $\|T_n\| \to \|T\|$. For example, let $T_n : \ell^2 \to \mathbb{C}$ be the linear operator given by $T_n(\{a_k\}) = a_n$. For each $a = \{a_k\}$ in ℓ^2 we have $\lim_{n \to \infty} T_n(a) = 0$, so $T \equiv 0$ is the pointwise limit of the operators T_n. However, $\|T_n - T\| = \|T_n\| = 1$.

We give some examples to illustrate applications of Theorems 3.11 and 3.12. The first uses the sequence space $c_0 = \{\{a_n\}_1^\infty : \lim_{n \to \infty} a_n = 0\}$, in the supremum norm. It is left to the reader (see Exercise 3.16) to show that c_0 is a closed subspace of ℓ^∞, hence is itself a Banach space.

Example 3.13. Suppose that $\{a_n\}_1^\infty$ is a sequence of complex numbers such that $\sum_1^\infty a_n b_n$ converges whenever $\{b_n\}_1^\infty$ is in c_0. We will show that $\sum_1^\infty |a_n| < \infty$. To see this, define $T_k : c_0 \to \mathbb{C}$ by

$$T_k(\{b_n\}) = \sum_{j=1}^k a_j b_j.$$

Each T_k is a bounded linear functional on c_0 with $\|T_k\| \leq \sum_{j=1}^k |a_j|$; the latter statement follows from the calculation

$$\left| \sum_{j=1}^k a_j b_j \right| \leq \sum_{j=1}^k |a_j b_j| \leq \left(\max_{1 \leq j \leq k} |b_j| \right) \sum_{j=1}^k |a_j| \leq \|\{b_n\}\|_\infty \sum_{j=1}^k |a_j|.$$

In fact, we have equality: $\|T_k\| = \sum_{j=1}^k |a_j|$. To see this, consider T_k acting on the unit vector in c_0

$$\left(\frac{\overline{a_1}}{|a_1|}, \frac{\overline{a_2}}{|a_2|}, \cdots, \frac{\overline{a_k}}{|a_k|}, 0 \cdots \right)$$

(with the obvious modifications if some $a_j = 0$), whose image under T_k is $\sum_{j=1}^k |a_j|$. We are given, then, that for each $b = \{b_n\} \in c_0$, $\lim_{k \to \infty} T_k(b)$ exists, since $T_k(b) = \sum_1^k a_j b_j$ and $\sum_1^\infty a_j b_j$ converges. In particular, $\sup_k |T_k(b)| < \infty$ for each $b \in c_0$. By Theorem 3.11, $\sup_k\{\|T_k\|\} < \infty$, and thus $\sum_{j=1}^\infty |a_j| < \infty$.

Example 3.14. Our next example is an application of Theorem 3.11 to a question about convergence of Fourier series. A continuous function f on the unit circle T has a Fourier series

$$\sum_{k=-\infty}^{\infty} a_k e^{ikx},$$

where

$$a_k = \hat{f}(k) = \frac{1}{2\pi} \int_{-\pi}^{\pi} f(t)e^{-ikt}\,dt$$

(that is, the Fourier expansion of f with respect to the orthonormal basis $\{e^{inx}\}_{-\infty}^{\infty}$ for $L^2(T, dx/(2\pi))$. We know that this series converges to f in the norm of $L^2(T)$, and so a subsequence converges pointwise almost everywhere; this is true for any f in $L^2(T)$, not just the continuous ones. But the general question of the point-wise convergence of the series is more delicate. In fact, for a period of time that stretched for 40 years, Riemann, Dirichlet, Weierstrass, and Dedekind all believed that the Fourier series of a continuous function converged pointwise everywhere (necessarily to the function value). The first counterexample was given in 1876 by DuBois-Reymond. Then the pendulum swung and for a while it was believed that the Fourier series of a continuous function could fail to converge at every point. In 1966 Lennart Carleson settled the matter definitively by proving the "Lusin con-jecture," which asserts that the Fourier series of any function in $L^2(T)$ (and thus, in particular, of any continuous function) converges pointwise almost everywhere. What we will do in this example is use the principle of uniform boundedness to show there exists an $f \in C(T)$ such that $s_n(f, 0)$ does not converge to $f(0)$, where $s_n(f, 0)$ denotes the symmetric partial sum $\sum_{k=-n}^{n} \hat{f}(k)e^{ikt}$ evaluated at 0. Indeed, we will show the existence of an $f \in C(T)$ so that the partial sums of the Fourier series of f at $t = 0$ are unbounded.

We begin with a calculation.

$$s_N(f, t) \equiv \sum_{k=-N}^{N} \hat{f}(k)e^{ikt} = \sum_{k=-N}^{N} \left(\int_{-\pi}^{\pi} f(x)e^{-ikx}\frac{dx}{2\pi} \right) e^{ikt}$$

$$= \int_{-\pi}^{\pi} f(x) \sum_{k=-N}^{N} e^{ik(t-x)}\frac{dx}{2\pi}$$

$$= \int_{-\pi}^{\pi} f(x)D_N(t-x)\frac{dx}{2\pi}$$

where $D_N(s) = \sum_{k=-N}^{N} e^{iks}$; this is the so-called *Dirichlet kernel*. The reader may recognize the last integral as the convolution $f * D_N(t)$ of f and D_N. Next we claim that

$$D_N(s) = \frac{\sin(N+\frac{1}{2})s}{\sin(\frac{s}{2})}$$

when $s \neq 0$, and $2N+1$ when $s = 0$. In the case $s = 0$ this is clear; otherwise write

$$\sum_{k=-N}^{N} e^{iks} = e^{-iNs} \sum_{k=0}^{2N} e^{iks} = e^{-iNs}\frac{1-e^{i(2N+1)s}}{1-e^{is}}.$$

Multiplying numerator and denominator by $e^{-is/2}$ and using the identity $e^{-iy} - e^{iy} = -2i\sin y$ gives the desired result. This kernel $D_N(s)$ is badly behaved in two respects: it is not positive and $\|D_N\|_1$ is not bounded. To see the latter, we have

$$\|D_N\|_1 = \int_{-\pi}^{\pi} \frac{|\sin(N+\frac{1}{2})s|}{|\sin(\frac{s}{2})|} \frac{ds}{2\pi} = \int_0^{\pi} \frac{|\sin(N+\frac{1}{2})s|}{|\sin(\frac{s}{2})|} \frac{ds}{\pi}$$

$$\geq \frac{2}{\pi} \int_0^{\pi} \frac{|\sin(N+\frac{1}{2})s|}{s} ds$$

$$= \frac{2}{\pi} \int_0^{\pi(N+\frac{1}{2})} |\sin u| \frac{du}{u},$$

where we have used the estimate $0 \leq \sin t \leq t$ for $0 \leq t \leq \pi/2$ and made the substitution $u = (N+(1/2))s$. Now

$$\frac{2}{\pi} \int_0^{\pi(N+\frac{1}{2})} |\sin u| \frac{du}{u} \geq \frac{2}{\pi} \sum_{k=1}^{N} \int_{(k-1)\pi}^{k\pi} |\sin u| \frac{1}{k\pi} du = \frac{4}{\pi^2} \sum_{k=1}^{N} \frac{1}{k} \geq \frac{4}{\pi^2} \log(N+1).$$

Now let us get set up to use the principle of uniform boundedness. Recalling that $C(T)$ is a Banach space in the supremum norm, define the linear functional $\Lambda_n : C(T) \to \mathbb{C}$ by $\Lambda_n(f) = s_n(f,0)$. Since

$$s_n(f,0) = \int_{-\pi}^{\pi} f(x)D_n(-x)\frac{dx}{2\pi}$$

we see that

$$|\Lambda_n(f)| = \left| \int_{-\pi}^{\pi} f(x)D_n(-x)\frac{dx}{2\pi} \right| \leq \int_{-\pi}^{\pi} |f(x)||D_n(-x)|\frac{dx}{2\pi} \leq \|f\|_\infty \|D_n\|_1$$

so that Λ_n is bounded with norm at most $\|D_n\|_1$. We claim that we actually have equality: $\|\Lambda_n\| = \|D_n\|_1$. To see this, fix n and let $g(x)$ be defined to be 1 if $D_n(x) > 0$, to be -1 if $D_n(x) < 0$ and 0 if $D_n(x) = 0$. We may then find continuous and piecewise linear functions $f_j(x)$ with $-1 \leq f_j(x) \leq 1$ for all x and $f_j \to g$ pointwise on $[-\pi, \pi]$ as $j \to \infty$. By the dominated convergence theorem

$$\lim_{j \to \infty} \int_{-\pi}^{\pi} f_j(x)D_n(-x)\frac{dx}{2\pi} = \int_{-\pi}^{\pi} g(x)D_n(-x)\frac{dx}{2\pi} = \int_{-\pi}^{\pi} |D_n(-x)|\frac{dx}{2\pi} = \|D_n\|_1.$$

Since $\|f_j\|_\infty \leq 1$ this shows that $\|\Lambda_n\| \geq \|D_n\|_1$ as desired.

We are finally ready to make our appeal to the principle of uniform boundedness. Either $\|\Lambda_n\| \leq M$ for some $M < \infty$ and for all n, or there exists $f \in C(T)$ such that $\sup_n |\Lambda_n(f)| = \infty$. Since $\|\Lambda_n\| = \|D_n\|_1 \to \infty$, the first alternative cannot hold and thus the second must. We obtain the existence of an $f \in C(T)$ such that

$$\sup_n |\Lambda_n(f)| = \sup_n |s_n(f,0)| = \infty,$$

and the Fourier series of f diverges at 0.

The next result is dual to the principle of uniform boundedness.

Theorem 3.15. *Let X be a normed linear space, and suppose A is a subset of X. If* $\sup\{|\varphi(x)| : x \in A\}$ *is finite for each fixed* φ *in* X^**, then A is bounded.*

Proof. Consider the natural map $\Phi : X \to X^{**}$ taking x to x^{**}. Note that $\Phi(A)$ is thus a collection of bounded linear functionals on X^*. Since X^* is a Banach space

$$\sup\{|\Phi(x)(\varphi)| : x \in A\} = \sup\{|\varphi(x)| : x \in A\} < \infty$$

for each $\varphi \in X^*$. By Theorem 3.11, applied to the linear maps

$$\mathscr{F} = \{\Phi(x) : x \in A\},$$

we must have

$$\sup\{\|\Phi(x)\| : x \in A\} < \infty.$$

However, we know that Φ is an isometry of X into X^{**}, so that $\|\Phi(x)\| = \|x\|$. Thus we conclude $\sup\{\|x\| : x \in A\} < \infty$; that is, A is bounded. □

Exercise 3.18 gives an application of Theorem 3.15.

3.3 Open Mapping and Closed Graph Theorems

The theorems of the title of this section are closely related; we will prove the open mapping theorem first, using the Baire category theorem, and then derive the closed graph theorem from it. An *open map* is one for which the image of every open set is open. The open mapping theorem concerns surjective maps in $\mathscr{B}(X,Y)$.

Theorem 3.16 (Open Mapping Theorem). *Suppose that X and Y are Banach spaces and that T is a bounded linear operator from X to Y. If T maps X onto Y, then T(G) is open in Y whenever G is open in X.*

Before we discuss the proof, let us give one important consequence. This is often called the inverse mapping theorem, and it is the third member of the triumvirate of results in this section.

Corollary 3.17 (Inverse Mapping Theorem). *Suppose X and Y are Banach spaces and* $T \in \mathscr{B}(X,Y)$ *is bijective. Its set-theoretic inverse* T^{-1} *is then a bounded linear operator from Y to X.*

Proof. We have already observed that T^{-1} exists as a linear map, so only boundedness of T^{-1} remains to be shown. By the open mapping theorem, T carries open sets to open sets. Now T^{-1} is bounded if and only if T^{-1} is continuous, and $T^{-1} : Y \to X$ is continuous if and only if $(T^{-1})^{-1}(G)$ is open in Y for every G that is open in X. But $(T^{-1})^{-1}(G) = T(G)$ and, by Theorem 3.16, $T(G)$ is open in Y for any open set G in X. Thus we conclude that T^{-1} is bounded, as desired. □

This answers our old question as to whether the set-theoretic invertibility of $T \in \mathscr{B}(X,Y)$ implies its operator-theoretic invertibility, i.e., the existence of an inverse in $\mathscr{B}(Y,X)$. We see the answer is yes, a result that Paul Halmos calls "one of the pleasantest and most useful facts about operator theory." The inverse mapping theorem was first proved by Banach in 1929. Our approach, using the open mapping theorem, is due to Schauder in 1930.

We turn next to the proof of the open mapping theorem. We will accomplish this by first proving the next result.

Theorem 3.18. *Suppose that X and Y are Banach spaces, and let B_X and B_Y denote the open unit balls, centered at 0, in X and Y, respectively. Suppose A is a bounded linear operator mapping X onto Y. There exists a positive constant δ such that $\delta B_Y \subseteq A(B_X)$; that is, given $y \in Y$ with $\|y\| < \delta$ there is $x \in X$ with $\|x\| < 1$ and $Ax = y$.*

Notice that the hypothesis that A is onto Y says that given any y in Y we may find an x in X with $Ax = y$; thus the significance of Theorem 3.18 is that we may control the norm of x in terms of the norm of y. Before we give the proof of Theorem 3.18, let us see that it will quickly yield the open mapping theorem.

Proof (Theorem 3.16). Let G be an open set in X and let x_0 be in G. We only need to show that $A(G)$ contains an open ball about Ax_0. To this end, translate G to obtain $G' \equiv G - x_0$. Since G' is an open set containing 0 we may find a positive number t with $tB_X \subseteq G'$. By Theorem 3.18 we have

$$A(G') \supseteq A(tB_X) = tA(B_X) \supseteq t\delta B_Y$$

for some positive constant δ. By linearity,

$$A(G) = A(G' + x_0) = Ax_0 + A(G') \supseteq Ax_0 + t\delta B_Y;$$

this last is the open ball centered at Ax_0 of radius $t\delta$. $\qquad\square$

To prove Theorem 3.18 we first give a lemma which is an approximate version of the theorem. It says that given $y \in Y$ we may get as close to y as desired by a vector of the form Ax for some x in X whose norm is controlled by the norm of y.

Lemma 3.19. *Suppose that X and Y are Banach spaces, and that A is a bounded linear operator mapping X onto Y. There is a positive number d with the following property: Given $\varepsilon > 0$ and $y \in Y$ there exists $x \in X$ such that $\|Ax - y\| < \varepsilon$ and $\|x\| < d^{-1}\|y\|$.*

Proof. Given $y \in Y$ there exists $\tilde{x}$ in X with $A\tilde{x} = y$, since A is surjective. This means

$$Y = \bigcup_{k=1}^{\infty} A(kB_X)$$

where B_X is the open unit ball in X. Since Y is a complete metric space, the Baire category theorem says that for some k, $A(kB_X)$ has closure with nonempty interior; say

$$\overline{A(kB_X)} \supseteq B(y_0, r)$$

for some $r > 0$ and $y_0 \in Y$. If $\|y\| < r$, then $y + y_0$ will be in $B(y_0, r)$ and hence in $\overline{A(kB_X)}$. Thus for any y in Y with $\|y\| < r$ we may find sequences $\{x'_n\}$ and $\{x''_n\}$ in kB_X such that $Ax'_n \to y_0$ and $Ax''_n \to y_0 + y$. Consider $x_n \equiv x''_n - x'_n$, and note that $Ax_n \to y$ and $\|x_n\| < 2k$.

The conclusion will follow from exploiting linearity. Let $z \neq 0$ be an arbitrary vector in Y, so that $(r/2)(z/\|z\|)$ is a vector in Y of norm less than r. By the first part of the proof we may find a sequence x_n in X with $\|x_n\| \leq 2k$ and $Ax_n \to (r/2)(z/\|z\|)$. Linearity says $A((2/r)\|z\|x_n) \to z$ where the norm of $(2/r)\|z\|x_n$ is less than $(4k/r)\|z\|$. This is the desired conclusion, with $d = r/(4k)$. $\qquad\square$

We can now prove Theorem 3.18 by an iterative use of this lemma. In the statement of the Lemma 3.19, we will refer to y as the *target vector* and ε as the *tolerance*.

Proof (Theorem 3.18). Let A, X, and Y be as in the statement of the theorem, and let d be as given by Lemma 3.19. Fix y in dB_Y, the open unit ball of radius d centered at 0 in Y. We apply the lemma, with target vector y and tolerance $\varepsilon = d/2$, to find $x_1 \in X$ of norm less than $(1/d)\|y\| < 1$ such that $\|y - Ax_1\| < d/2$. Apply the lemma again, this time with target $y - Ax_1$ and tolerance $\varepsilon = d/4$ to find x_2 in X with

$$\|(y - Ax_1) - Ax_2\| < \frac{d}{4}$$

and

$$\|x_2\| < \frac{1}{d}\|y - Ax_1\| < \frac{1}{2}.$$

Continue inductively, so that if we have determined $x_1, x_2, \ldots, x_n$ with

$$\|y - Ax_1 - Ax_2 - \cdots - Ax_n\| < \frac{d}{2^n}$$

and

$$\|x_k\| < \frac{1}{2^{k-1}} \text{ for } k = 1, 2, \ldots, n,$$

then at the next step we apply the lemma with target $y - Ax_1 - \cdots - Ax_n$ and tolerance $d/2^{n+1}$ to select x_{n+1} so that

$$\|y - Ax_1 - \cdots - Ax_n - Ax_{n+1}\| < \frac{d}{2^{n+1}}$$

and

$$\|x_{n+1}\| < \frac{1}{d}\|y - Ax_1 - \cdots - Ax_n\| < \frac{1}{2^n}.$$

For each positive integer n, define $v_n = x_1 + x_2 + \cdots + x_n$ and observe that $\{v_n\}$ is a Cauchy sequence in X: When $m > n$,

$$\|v_m - v_n\| = \|x_{n+1} + \cdots + x_m\| < \sum_{n+1}^{m} \frac{1}{2^k} \to 0$$

as $n, m \to \infty$. By completeness, there is an $x \in X$ with $v_n \to x$. Moreover,

$$\|x\| \le \sum_{1}^{\infty} \|x_k\| < \sum_{1}^{\infty} \frac{1}{2^{k-1}} = 2$$

so that x is in $2B_X$. Since $\|y - Av_n\| < d/2^n$ we have $Av_n \to y$ as $n \to \infty$. By continuity of A, $Av_n \to Ax$, so that $y = Ax$. Recalling that y was arbitrary in dB_Y and x is in $2B_X$, we see that we have proved $A(2B_X) \supseteq dB_Y$, and by linearity $A(B_X) \supseteq (d/2)B_Y$. This gives Theorem 3.18, with $\delta = d/2$. $\qquad\square$

Definition 3.20. When X and Y are normed linear spaces and $T : X \to Y$ is a linear map, the *graph* of T, denoted graph(T), is $\{(x, Tx) : x \in X\}$. Note that the graph of T is a subset of $X \times Y$.

The product $X \times Y$ is a vector space under coordinatewise operations. We can put a norm on $X \times Y$ (the "one-norm") by $\|(x,y)\| = \|x\|_X + \|y\|_Y$. It is not hard to show that when X and Y are Banach spaces, then $X \times Y$ in the one-norm is also a Banach space; see Exercise 3.19. Notice also that the graph of T is a (not necessarily closed) subspace of $X \times Y$.

The next result, called the closed graph theorem, gives a new way to see if a linear map between Banach spaces is bounded.

Theorem 3.21 (Closed Graph Theorem). *If X and Y are Banach spaces and $T : X \to Y$ is linear, then T is bounded if and only if graph(T) is closed in $X \times Y$.*

Before we give the proof, we make a few observations. The hypothesis that graph(T) is closed can be reformulated as "whenever (x_n, Tx_n) converges to (x,y) in $X \times Y$, then we must have $y = Tx$". The "only if" direction of Theorem 3.21 is trivial: If T is bounded and if $(x_n, Tx_n) \in$ graph(T) satisfies $(x_n, Tx_n) \to (x,y)$, we have $\|x_n - x\|_X \to 0$ and $\|Tx_n - y\|_Y \to 0$. Continuity of T implies that $\|Tx_n - Tx\|_Y \to 0$, so that $Tx = y$.

Proof (Theorem 3.21). Only the "if" direction needs proof. If the graph of T is closed, then it is a closed subspace of the Banach space $X \times Y$ (in the one-norm), and thus is itself a Banach space. Consider the continuous linear maps $P_1 : $ graph$(T) \to X$ and $P_2 : $ graph$(T) \to Y$ defined by $P_1(x, Tx) = x$ and $P_2(x, Tx) = Tx$. The map P_1 is bijective, and thus, by the inverse mapping theorem, P_1^{-1} is a continuous linear map of X onto graph(T). Since we can write $T = P_2 \circ P_1^{-1}$, we see that T is continuous. $\qquad\square$

Let us think about what this result actually does for us. If X and Y are Banach spaces and $T : X \to Y$ is linear, to show that T is continuous from the definition, we assume that $x_n \to x$, and then must show both that Tx_n converges, and that its limit is Tx. By contrast, with the closed graph theorem at our disposal, to show that T is continuous, we may assume both

$$x_n \to x \quad \text{and} \quad Tx_n \to y;$$

our task is then simply to show that $y = Tx$. In fact, its even a bit simpler. By linearity, we need only show that whenever $x_n \to 0$ and $Tx_n \to y$, then $y = 0$; see Exercise 3.21.

As an application of the closed graph theorem, we next prove the two-norm theorem.

Theorem 3.22 (Two-Norm Theorem). *Suppose X is a normed linear space with two norms, $\| \cdot \|_1$ and $\| \cdot \|_2$, each of which make X into a Banach space. If there exists a finite constant M such that*

$$\|x\|_1 \leq M \|x\|_2$$

for all $x \in X$, then there exists a finite constant K such that $\|x\|_2 \leq K \|x\|_1$ for all $x \in X$.

Proof. Let $I : (X, \| \cdot \|_1) \to (X, \| \cdot \|_2)$ be the identity map. Clearly I is linear, and we want to show that it is bounded. To do this we will apply the closed graph theorem. Suppose that $x_n \to x$ in $(X, \| \cdot \|_1)$ and that $I(x_n) = x_n \to y$ in $(X, \| \cdot \|_2)$. Our goal is to show that $y = Ix = x$. For each n,

$$\|I(x) - y\|_1 = \|x - y\|_1 \leq \|x - x_n\|_1 + \|x_n - y\|_1 \leq \|x - x_n\|_1 + M\|x_n - y\|_2,$$

which tends to 0 as $n \to \infty$. Hence $x = y$, and by Theorem 3.21 we conclude that I is bounded. $\qquad \square$

When a pair of norms $\| \cdot \|_1$ and $\| \cdot \|_2$ satisfy both $\|x\|_1 \leq M\|x\|_2$ and $\|x\|_2 \leq K\|x\|_1$ for finite constants M and K, we say the norms are *equivalent*. Note that equivalent norms will induce the same topology on the underlying space, since an open set in one norm is also an open set in the other norm.

As an application of the two-norm theorem, we will show that $C[0,1]$ in the L^1 norm $\|f\|_1 = \int_0^1 |f| dx$ is not a Banach space. We know that $(C[0,1], \| \cdot \|_\infty)$ is a Banach space, and it is trivial that $\|f\|_1 \leq \|f\|_\infty$ holds for all f in $C[0,1]$. If $(C[0,1], \| \cdot \|_1)$ were a Banach space, the two-norm theorem would say that $\|f\|_\infty \leq K\|f\|_1$ for some finite constant K and all $f \in C[0,1]$. The reader can easily construct piecewise linear functions f_n in $C[0,1]$ with $\|f_n\|_1 = 1$ and $\|f_n\|_\infty = n$; for example, let $f_n(x)$ be $n - (n^2/2)x$ for $0 \leq x \leq 2/n$ and 0 elsewhere.

3.4 Quotient Spaces

As an application of the results in the previous section, we consider the notion of a *quotient space* of a Banach space. Suppose that X is a Banach space, and M is a closed subspace of X. Define an equivalence relation on X by decreeing that $x \cong y$ if and only if $x - y$ is in M; this is easily verified to be an equivalence relation. Denote the set of equivalence classes by X/M; that is, X/M is the set of cosets $x + M$ where $x_1 + M = x_2 + M$ if and only if $x_1 - x_2$ is in M. Define addition and scalar multiplication on X/M by

$$(x_1 + M) + (x_2 + M) = x_1 + x_2 + M$$

and

$$\alpha(x + M) = \alpha x + M.$$

With these definitions, X/M becomes a vector space with zero vector $0 + M = M$. Put what will be a norm on X/M by setting

$$\|x + M\| = \inf\{\|x + m\| : m \in M\} = \inf\{\|x - m\| : m \in M\},$$

so that $\|x + M\|$ can be thought of as the distance from x to M. Note that $\|x + M\| = 0$ if and only if x is in M; we are using the hypothesis that M is closed. In Exercise 3.25 you are asked to show that this is indeed a norm on X/M, and that X/M is complete. The map $\Pi : X \to X/M$ which sends x to $x + M$ is called the natural, or quotient, map. It is linear, and since

$$\|\Pi(x)\| = \|x + M\| \leq \|x\|,$$

it is bounded.

As a particular application of these ideas, consider a bounded linear operator $T : X \to Y$, where X and Y are Banach spaces and let $M = \ker T$, a closed subspace in X. Consider the quotient $X/M = X/\ker T$. If T is surjective, we claim that Y and $X/\ker T$ are isomorphic; that is, there is a bijective bounded linear operator from one onto the other, with bounded inverse. To see this, set

$$A : X/\ker T \to Y$$

by

$$A(x + \ker T) = Tx.$$

It is easy to verify that A is well-defined, linear, and bounded; see Exercise 3.26. Moreover, if $A(x_1 + \ker \mathrm{T}) = A(x_2 + \ker \mathrm{T})$, then $Tx_1 = Tx_2$ and $x_1 - x_2$ is in $\ker T$. This shows that A is one-to-one. To see that A is onto Y, let y be in Y and find x with $Tx = y$, so that $A(x + \ker T) = y$. Once we know that A is bijective and linear, the inverse mapping theorem guarantees that A has a bounded inverse, and therefore $X/\ker T$ is isomorphic to Y.

As a special case of this, suppose T is a bounded linear functional on a Banach space X that is not the zero functional. In this case, T is automatically surjective, so we conclude that $X/\ker T$ is isomorphic to $\mathbb{C}$. This result will be important to us in Chapter 5.

It is interesting to look at $\mathcal{H}/M$ in the special case that M is a closed subspace of a Hilbert space $\mathcal{H}$. Exercise 3.26 asks you to show that in this case the quotient map gives an isometric isomorphism of $M^\perp$ onto $\mathcal{H}/M$. The philosophy is then that X/M acts as a substitute for $M^\perp$ in the Banach, non-Hilbert space setting.

3.5 Banach and the Scottish Café

By now we have seen evidence of Banach's central role in the development of functional analysis in roughly the period from 1920 (when he completed his doctoral thesis) to his death in 1945, at which point the theory of linear operators on Banach and Hilbert spaces had reached a level of maturity. Bourbaki (see the footnote in Section 1.4) makes the following comment:

> The publication of Banach's treatise "Opérations Linéaires" marks, one could say, the beginning of the adult age for the theory of normed spaces.... As it happened, the work had considerable success...([6], p. 347).

Here we will say a bit more about his life and mathematical colleagues in Poland. Much more information can be found in R. Kaluza's biography of Banach [26].

Banach was a protege of Steinhaus, who was only a few years older than Banach, and together they founded the Polish school of functional analysis, often referred to as the Lwów school, which flourished during the period between World War I and World War II. For a time the Café Szkocka ("Scottish Café") in Lwów served as a prime location for collaborative work and discussion between the members of this Lwów school. Meetings of the Polish Mathematical Society held at the Mathematics Department at the University of Lwów were followed by discussions, first at the nearby Café Roma, and then later next door at the Café Szkocka, which evidently offered Banach a more congenial credit situation. Eventually this became the site of near daily meetings, and a notebook purchased by Banach's wife became a repository for problems posed by mathematicians working in the Scottish Café (prior to this purchase, problems and work simply got written on the marble tabletops of the café, to be erased at closing time by the janitor). The first entry in the "Scottish Problem Book" as it came to be known, was made in July 1935 and the 193rd—and last—entry in May 1941. Space was left after each problem for any forthcoming solution to be added later. The book was kept at the café, to be produced by a waiter or cashier when called for by Banach or one of his colleagues. Visitors added to the problem book too, and one can see hints of the larger political landscape in this; for example, Russian names appear among the contributors after 1939, when the Soviet Union occupied Lwów. Prizes were offered for solutions to some of the problems, many of these involved alcohol: "two small beers," "a flask of brandy," or "a bottle of wine," while others were more unusual: "a live goose," or hinted at deprivations brought on by world events, "a kilo of bacon." As World War II loomed, there were concerns for the safety of the book and various suggestions were made for its safe-keeping. The book did survive the war, and the original remains in the possession of Banach's son. A copy of the book was published by Birkhäuser in 1981 [31], along with considerable commentary on the problems. Many parts of mathematics—not just functional analysis—are represented.

When Germany invaded the Soviet Union and then entered Lwów in 1941, Banach faced danger, both as a member of the "Polish intellectual elite" and for having had good relations with the Soviets during the previous period. That he escaped death when many Polish scholars were executed was perhaps due to his employment as a "lice-feeder" in the Weigl Institute in Lwów. Run by the Polish biologist Rudolf Stefan Weigl, the Institute produced a typhus vaccine by a process which required daily feedings of lice on the blood of human hosts. Weigl was able to offer some measure of protection for employees of the Institute, many of whom were university professors, from arrest and deportation to concentration camps. This was both because the work of the institute was considered a priority by the Germans, and at the same time, the Gestapo was disinclined to interfere with Institute employees for fear they could be carrying typhus-infected lice. Institute employees carried special identity papers which included warnings of this risk. Many of the feeders in the particular unit in which Banach worked were also mathematicians, and lively mathematical discussions continued during the time when the lice were feeding. Banach worked in the Weigl Institute from the fall of 1941 until Soviet troops reentered Lwów in July 1944. An underground university, formed under cover of the Institute, also came into existence during this period. Banach taught in this university, and according to a reminiscence of Banach written by Steinhaus [44], one student received a doctorate under his direction during this time. Although Banach survived the period of Nazi occupation of Lwów, he suffered under the harsh conditions of the time, with illness and malnutrition, and died at the young age of 53 in 1945 of lung cancer.

3.6 Exercises

3.1. Use Zorn's lemma to prove the following: If E is an orthonormal set in a Hilbert space $\mathscr{H}$, then $\mathscr{H}$ has an (orthonormal) basis containing E. In particular, every Hilbert space has an orthonormal basis. A similar Zorn's lemma argument shows that every vector space has a Hamel basis.

3.2. We introduce some terminology for the purpose of this problem: If X is either a real or complex vector space (meaning that the scalars used in scalar multiplication are real, or, respectively, complex), we say that a real-valued φ is a *real-linear functional* if $\varphi(x+y) = \varphi(x) + \varphi(y)$ and $\varphi(\alpha x) = \alpha\varphi(x)$ holds for all $x, y \in X$ and α real. For X a complex vector space, we say that (a complex-valued) φ is a *complex-linear functional* if these relationships hold for all $x, y \in X$ and α complex.

(a) Show that for any complex number z, $z = \operatorname{Re} z - i\operatorname{Re}(iz)$.
(b) Suppose X is a complex vector space and φ is a complex-linear functional on X. Define $u : X \to \mathbb{R}$ by $u(x) = \operatorname{Re} \varphi(x)$. Show that $\varphi(x) = u(x) - iu(ix)$ for all $x \in X$.
(c) Suppose u is a real-linear functional on X. Define $\varphi : X \to \mathbb{C}$ by $\varphi(x) = u(x) - iu(ix)$. Show that φ is a complex-linear functional on X. (Hint: Check the condition $\varphi(\alpha x) = \alpha\varphi(x)$ first for α real, then for $\alpha = i$, then for α complex.)
(d) Now suppose X is a normed linear space. For φ and u related as above, show that $\|\varphi\| = \|u\|$.

3.3. (Complex Hahn–Banach). Suppose Y is a subspace of a complex normed linear space X and $\varphi : Y \to \mathbb{C}$ is a bounded, complex-linear functional on Y. Show that φ extends to a bounded complex-linear functional Φ on X with $\|\varphi\| = \|\Phi\|$. Hints: Most of the work is done by the previous problem. Let $u = \operatorname{Re} \varphi$ and use the real Hahn–Banach theorem to extend u to U on all of X. Define $\Phi(x) = U(x) - iU(ix)$ and check that Φ has the desired properties. This correspondence between U and Φ was observed by Löwig in 1934.

3.4. Let $L_{\mathbb{R}}^{\infty}$ be the space of real-valued essentially bounded functions on $[0,1]$ with respect to Lebesgue measure. Let M be the subspace of constant functions. Define $f : M \to \mathbb{R}$ by $f(c) = c$, where on the left hand side, c denotes the constant function with value c. Let $g_0(x) = x$, and set $N \equiv \{c + tg_0 : c \in M, t \in \mathbb{R}\}$. The proof of the one-step extension process in the Hahn–Banach theorem tells you how to find all linear $F : N \to \mathbb{R}$ so that F extends f and $\|F\| = \|f\|$. Find all such F.

3.5. For $1 \le p < \infty$ it is a fact that the dual space to $L^p(X, \mu)$, where (X, μ) is a σ-finite measure space, is $L^q(X, \mu)$, $1/p + 1/q = 1$, in the following sense: Given $g \in L^q(X, \mu)$, define $\Lambda_g(f) = \int_X fg\,d\mu$. This is a bounded linear functional on $L^p(X, \mu)$, and $\|\Lambda_g\| = \|g\|_q$. Conversely, every bounded linear functional on $L^p(X, \mu)$ has this form. Using this and the Hahn–Banach theorem, show

$$\|f\|_p = \sup\left\{ \left| \int_X fg\,d\mu \right| : g \in L^q(X), \|g\|_q = 1 \right\}$$

for all $f \in L^p(X)$.

3.6. Suppose that (X, μ) is a positive, σ-finite measure space. Let $\{g_n\}$ be a sequence in $L^3(X, \mu)$ such that $\sup_n \|g_n\| = \infty$. Prove there exists a function $f \in L^{3/2}(X, \mu)$ such that $\sup_n |\int fg_n\,d\mu| = \infty$. You may use the fact that the dual space of $L^p(\mu), 1 \le p < \infty$, is $L^q(\mu)$, $1/p + 1/q = 1$ in the sense that is described in the previous exercise.

3.7. Let X be a compact Hausdorff space. A *positive linear functional* on $C(X)$ is a (bounded) linear functional Λ with the additional property that $\Lambda(f) \ge 0$ whenever $f \ge 0$ on X. Show that point evaluation at $x_0 \in X$ is a positive linear functional for each $x_0 \in X$.

There is a representation theorem for the positive linear functionals on $C(X)$ which says that for each positive linear functional Λ there is a unique positive, finite, regular Borel measure μ on X with

$$\Lambda(f) = \int_X f d\mu.$$

(See Section A.5 in the Appendix for further discussion). If Λ is point evaluation at x_0, what is the corresponding measure μ?

3.8. Prove Corollary 3.6.

3.9. Let $\ell_{\mathbb{R}}^\infty$ denote the space of bounded sequences with real entries, in the supremum norm. Consider the operator T defined on $\ell_{\mathbb{R}}^\infty$ by $T(x_1, x_2, \ldots) = (x_2, x_3, \ldots)$; this is clearly bounded. Let $M = \operatorname{ran}(T - I)$, a subspace of $\ell_{\mathbb{R}}^\infty$. Set $e = (1, 1, 1, \ldots) \in \ell_{\mathbb{R}}^\infty$, and note that since $0 \in M$, dist $(e, M) \le \|e\|_\infty = 1$.

(a) Show that in fact dist $(e, M) = 1$. Hint: Argue by contradiction.
(b) Show that there exists a bounded linear functional $\varphi : \ell_{\mathbb{R}}^\infty \to \mathbb{R}$ with $\|\varphi\| = 1$, $\varphi(e) = 1$, and $\varphi(T\{x_n\}) = \varphi(\{x_n\})$ for every $\{x_n\}$ in $\ell_{\mathbb{R}}^\infty$.
(c) Let c be real and $s > 0$. Consider a sequence $\{x_n\}$ with

$$c - s \le x_n \le c + s$$

for all $n \in \mathbb{N}$. Show that

$$c - s \le \varphi(\{x_n\}) \le c + s.$$

(d) Show that for any $k = 0, 1, 2, \ldots$ and $\{x_n\}$ in $\ell_{\mathbb{R}}^\infty$,

$$\varphi(T^{k+1}\{x_n\}) = \varphi(T^k\{x_n\}).$$

Conclude that
$$\varphi(\{x_1, x_2, \ldots\}) = \varphi(\{x_N, x_{N+1}, \ldots\})$$

for every $N \in \mathbb{N}$.
(e) Show that if $\{x_n\} \in \ell_{\mathbb{R}}^\infty$, then

$$\liminf_{n \to \infty} x_n \le \varphi(\{x_n\}) \le \limsup_{n \to \infty} x_n.$$

Such a linear functional is called a *Banach limit*. Note that if $\lim_{n \to \infty} x_n$ exists, it must be $\varphi(\{x_n\})$.

3.10. Given a normed linear space X and a (not necessarily closed) subspace M of X, define
$$M^\perp = \{\varphi \in X^* : \varphi(x) = 0 \text{ for all } x \in M\},$$

the bounded linear functionals that vanish on M. Call this the *annihilator* of M, and note that the notation is consistent with our earlier usage in the context of Hilbert spaces. Furthermore, if N is a (again, not necessarily closed) subspace of X^*, define

$$^\perp N = \{x \in X : \varphi(x) = 0 \text{ for all } \varphi \in N\},$$

so that $^\perp N$ is the set of common zeros of the bounded linear functionals in N. Show that for any subspace M of X,

$$^\perp(M^\perp) = \text{closure } M.$$

3.11. Show that if X is a Banach space that is not reflexive, then X^* is also not reflexive. Hint: Find a nonzero bounded linear functional on X^{**} which is 0 on $\{x^{**} : x \in X\}$.

The converse statement is also true; see p. 132 in [8].

3.12. Use the Baire category theorem to show that no infinite-dimensional Hilbert space can have a countable Hamel basis.

3.13. The point of this problem is to show that Theorem 3.11 may fail in a normed linear space that is not a Banach space. Let F be the set of "eventually zero" sequences, in the supremum norm; this means that a sequence $\{a_n\} \in \ell^\infty$ belongs to F if there is an N with $a_n = 0$ for all $n \geq N$. Define linear maps $T_n : F \to \mathbb{C}$ by

$$T_n(\{a_k\}) = \sum_{k=1}^{n} a_k.$$

Show that each T_n is linear and bounded and for any fixed sequence $x = \{a_k\}$ in F, $\sup\{|T_n(x)| : n = 1, 2, 3, \ldots\}$ is finite. Is $\sup\{\|T_n\| : n = 1, 2, 3 \ldots\} < \infty$?

3.14. Let $\mathscr{H}$ be a Hilbert space. Let $\{x_n\}$ be a sequence in $\mathscr{H}$ with the property that $\langle x, x_n \rangle \to 0$ as $n \to \infty$ for each vector $x \in \mathscr{H}$. Show that $\sup\{\|x_n\| : n = 1, 2, 3, \ldots\} < \infty$.

3.15. In this problem we outline the "gliding hump" technique as originally used by Banach and Steinhaus to prove the uniform boundedness principle. This outline is taken from [34]. We keep the notation as in Theorem 3.11. Denote

$$\sup\{\|Tx\| : T \in \mathscr{F}\} = m(x)$$

so that $m(x) < \infty$ for each $x \in X$. Assume, for a contradiction, that $\sup\{\|T\| : T \in \mathscr{F}\} = \infty$.

(a) Show that by an inductive construction we may find $T_1, T_2, \ldots$ in $\mathscr{F}$ and $x_1, x_2, \ldots$ in X with

$$\frac{1}{4} \frac{1}{3^n} \|T_n\| \geq \sum_{k<n} m(x_k) + n$$

$$\|x_n\| \leq \frac{1}{3^n}$$

and

$$\|T_n x_n\| \geq \frac{3}{4} \frac{1}{3^n} \|T_n\|.$$

(Determine the T_j and x_j alternately.)

(b) Set $x = \sum_{k=1}^{\infty} x_k$ so that $T_n x = \sum_{k<n} T_n x_k + T_n x_n + \sum_{k>n} T_n x_k$; the middle term being the "gliding hump." Observe that

$$\left\| \sum_{k<n} T_n x_k \right\| \leq \sum_{k<n} m(x_k)$$

and

$$\left\| \sum_{k>n} T_n x_k \right\| \leq \sum_{k>n} \frac{1}{3^k} \|T_n\| \leq \frac{1}{2 \cdot 3^n} \|T_n\|.$$

Argue that $\|T_n x\| \geq n$, a contradiction.

3.16. Show that

$$c_0 = \{\{a_n\}_1^{\infty} : \lim_{n \to \infty} a_n = 0\}$$

is a closed subspace of ℓ^{∞}, in the supremum norm.

3.17. For any $a = \{a_n\}$ in ℓ^1, define a linear functional φ_a on c_0 by

$$\varphi_a(\{x_n\}) = \sum_{n=1}^{\infty} a_n x_n.$$

Show that the map $a \to \varphi_a$ is an isometric isomorphism of ℓ^1 onto $(c_0)^*$; that is, $(c_0)^* \cong \ell^1$.

3.18. Let X and Y be normed linear spaces and suppose $T : X \to Y$ is linear. Show that T is continuous if $\varphi \circ T$ is continuous for all φ in X^*.

3.19. Suppose that X and Y are Banach spaces.

(a) Show that $X \times Y$ in the one-norm

$$\|(x,y)\| \equiv \|x\|_X + \|y\|_Y$$

is a Banach space.

(b) Is $X \times Y$ a Banach space in the norm

$$\|(x,y)\|_{\infty} \equiv \max(\|x\|_X, \|y\|_Y)?$$

3.20. Let c denote the linear subspace of ℓ^{∞} consisting of all sequences $x = \{x_n\}_1^{\infty}$ for which $\lim_{n \to \infty} x_n$ exists.

(a) Let $e = (1,1,1,\dots) \in c$. Show that

$$c = \{x + \alpha e : x \in c_0 \text{ and } \alpha \in \mathbb{C}\}.$$

(b) Argue that the formula $\varphi_{\infty}(\{x_n\}) = \lim_{n \to \infty} x_n$ defines a bounded linear functional on c, where c is equipped with the supremum norm.

(c) Show that c is a closed subspace of ℓ^∞.

(d) Given $b = \{b_n\}$ in ℓ^1 and $\gamma \in \mathbb{C}$, consider the linear functional defined on c by

$$\psi_{b,\gamma}(\{x_n\}) = \sum_{n=1}^{\infty} b_n x_n + \gamma \lim_{n\to\infty} x_n.$$

Show that the map $(b, \gamma) \to \psi_{b,\gamma}$ is an isometric isomorphism of $\ell^1 \times \mathbb{C}$, equipped with the norm $\|(b, \gamma)\| = \|b\|_1 + |\gamma|$, onto c^*.

3.21. Suppose X, Y are Banach spaces and $T : X \to Y$ is linear. Suppose further that whenever $x_n \to 0$ and $Tx_n \to y$ then $y = 0$. Show that T is continuous.

3.22. Suppose that $\varphi : \mathbb{D} \to \mathbb{D}$ is an analytic function (where $\mathbb{D}$ is the open unit disk) with the property that $f \in L_a^2(\mathbb{D})$ implies $f \circ \varphi \in L_a^2(\mathbb{D})$. Define $C_\varphi : L_a^2(\mathbb{D}) \to L_a^2(\mathbb{D})$ by $C_\varphi(f) = f \circ \varphi$. Show that the composition operator C_φ is a bounded linear operator on $L_a^2(\mathbb{D})$.

3.23. Let $X = C[0, 1]$ in the supremum norm and let

$$Y = C^1[0, 1] \equiv \{f \in C[0, 1] : f' \text{ exists and is continuous on } [0, 1]\}.$$

Give Y the supremum norm also. Define $T : Y \to X$ by $Tf = f'$. Clearly T is linear.

(a) Show that if $f_n \to f$ and $Tf_n \to g$, then $g = Tf$. (Hint: you need only show $g(x) = f'(x)$ for all $x \in [0, 1]$. Use the fundamental theorem of calculus).

(b) Show that T is not bounded.

(c) Why doesn't this contradict the closed graph theorem?

3.24. Use the closed graph theorem to show that the operator

$$Bf = \frac{f - f(0)}{z}$$

is a bounded linear operator on $L_a^2(\mathbb{D})$.

3.25. Show that the quotient X/M of a Banach space X by a closed subspace M is a Banach space. (Begin by showing that

$$\|x + M\| \equiv \inf\{\|x + m\| : m \in M\}$$

is a norm on X/M.)

3.26. (a) Let X and Y be Banach spaces and $T : X \to Y$ be a bounded linear operator. Show that

$$A : X/\ker T \to Y$$

given by $A(x + \ker T) = Tx$ is a well-defined, one-to-one, bounded linear operator.

(b) Suppose $T : X \to \mathbb{C}$ is a bounded linear functional, not identically 0, where X is a Banach space. Show that T must be surjective and conclude $X/\ker T$ is isomorphic to $\mathbb{C}$.

(c) Suppose $\mathcal{H}$ is a Hilbert space and M is a closed subspace of $\mathcal{H}$. Use the projection theorem to show that the quotient map $\Pi : \mathcal{H} \to \mathcal{H}/M$ gives an isometric isomorphism of $M^{\perp}$ onto $\mathcal{H}/M$.

3.27. Suppose that T is in $\mathcal{B}(\mathcal{H})$ for some Hilbert space $\mathcal{H}$ and that T has closed range. Show there exists $c > 0$ such that

$$\|Th\| \geq c\|h\|$$

for all $h \in (\ker T)^{\perp}$.

3.28. Give an example of a diagonal operator $T : \mathcal{H} \to \mathcal{H}$ whose range is not closed.

3.29. Let M be a closed subspace of a Banach space X.

(a) Show that the map defined on $X^*/M^{\perp}$, the quotient of X^* by the annihilator of M (see Exercise 3.10 for the definition), sending $\varphi + M^{\perp}$ to $\varphi|_M$ (the restriction of φ to M) is a well-defined, linear, isometric map of $X^*/M^{\perp}$ onto M^*. (In short, $X^*/M^{\perp} \cong M^*$).

(b) Show that the map from $(X/M)^*$ to $M^{\perp}$ which sends φ in $(X/M)^*$ to $\varphi \circ \Pi$, where Π is the quotient map from X to X/M, is a well-defined, linear isometry of $(X/M)^*$ onto $M^{\perp}$. (In short, $(X/M)^* \cong M^{\perp}$.)

3.30. Suppose that X is a functional Banach space (as defined in Section 1.4) of functions defined on a set S, and g is a scalar-valued function on S with the property that $f \in X$ implies $fg \in X$. Define $M_g : X \to X$ by $M_g f = fg$.

(a) Show that M_g is continuous and that g must be bounded.
(b) Show that $\sup\{|g(s)| : s \in S\} \leq \|M_g\|$. Give an example to show that this inequality may be strict.

3.31. Suppose that X is a functional Banach space over a set S, and that each function in X is bounded. Show that $\sup_{s \in S} \|e_s\| < \infty$, where e_s denotes the functional of evaluation at s.

3.32. Suppose that A is a linear map from a Hilbert space $\mathcal{H}$ into itself that satisfies $\langle x, Ay \rangle = \langle Ax, y \rangle$ for all x, y in $\mathcal{H}$. Show that A is bounded.

3.33. This problem outlines a proof of the statement: If X is a Banach space and $T \in \mathcal{B}(X)$ is such that X/TX, as a vector space, is finite-dimensional, then TX is closed.

(a) Argue that since the map $A : X/\ker T \to X$ defined by $A(x + \ker T) = Tx$ is one-to-one and has the same range as T, we may assume without loss of generality that T is one-to-one.

(b) Suppose that X/TX has dimension 1, so that there exists $y \in X$ such that $X = \{Tx + \alpha y : x \in X, \alpha \in \mathbb{C}\}$. Show that the map S defined on $X \times \mathbb{C}$ in the one-norm by $S(x, \alpha) = Tx + \alpha y$ is continuous and bijective. Use this to show that T is bounded below and thus has closed range.

(c) Prove the full result.

3.34. This problem outlines one way to show that the Banach space c_0 (as defined in Exercise 3.16) is not a reflexive Banach space.

(a) Show that $\varphi : c_0 \rightarrow \mathbb{C}$ defined by

$$\varphi(\{a_n\}) = \sum_{n=1}^{\infty} \frac{a_n}{n!}$$

is a bounded linear functional on c_0 and

$$\|\varphi\| = \sum_{n=1}^{\infty} \frac{1}{n!}.$$

(b) Show that for every $\{a_n\}$ in c_0 with $\|\{a_n\}\| = 1$,

$$|\varphi(\{a_n\})| < \sum_{n=1}^{\infty} \frac{1}{n!}.$$

(c) Conclude that c_0 is not reflexive.

3.35. Recall the notion of the strong operator topology from Exercise 2.23 in Chapter 2.

(a) Consider a sequence $\{T_n\}$ of bounded linear operators on a Hilbert space $\mathcal{H}$. Suppose that for each $h \in \mathcal{H}$, $\{T_n h\}$ is a Cauchy sequence in $\mathcal{H}$. Show that there exists $T \in \mathcal{B}(\mathcal{H})$ such that $T_n \rightarrow T$ (SOT). This result is sometimes phrased as "$\mathcal{B}(\mathcal{H})$ is sequentially complete in the strong operator topology."

(b) Suppose that $\{T_n\}$ is a sequence of operators in $\mathcal{B}(\mathcal{H})$ and suppose further that for each h, g in $\mathcal{H}$, $\langle T_n h, g \rangle$ converges as $n \rightarrow \infty$. Show that there exists T in $\mathcal{B}(\mathcal{H})$ such that $T_n \rightarrow T$ (WOT). Hints: Show first that for each $h \in \mathcal{H}$, $\sup_n \|T_n h\| < \infty$, by considering the family of bounded linear functionals $\langle \cdot, T_n h \rangle$, and then argue that $\sup_n \|T_n\| < \infty$. If $S(h, g) \equiv \lim_{n \to \infty} \langle T_n h, g \rangle$, then S is a bounded sesquilinear form.

3.36. A sequence $\{h_n\}$ of vectors in a Hilbert space $\mathcal{H}$ is said to be a *Bessel sequence* if

$$\sum_{n=1}^{\infty} |\langle h, h_n \rangle|^2 < \infty$$

for every $h \in \mathcal{H}$. A sequence $\{g_n\}$ is said to be a *Riesz–Fischer sequence* if given any $\{c_n\} \in \ell^2$ there exists (at least one) vector $g \in \mathcal{H}$ such that

$$\langle g, g_n \rangle = c_n \text{ for all } n. \tag{3.5}$$

Note that an orthonormal basis is both a Bessel sequence and a Riesz–Fischer sequence.

(a) Show that if $\{h_n\}$ is a Bessel sequence, then there exists $M < \infty$ so that

$$\sum_{n=1}^{\infty} |\langle h, h_n \rangle|^2 \leq M\|h\|^2$$

for all $h \in \mathcal{H}$. Hint: Apply the closed graph theorem to the map $S : \mathcal{H} \to \ell^2$ defined by $Sh = \{\langle h, h_n \rangle\}$.

(b) Show that if $\{g_n\}$ is a Riesz–Fischer sequence, there exists $m > 0$ such that given $\{c_n\} \in \ell^2$, the equations in (3.5) hold for at least one solution g satisfying

$$m\|g\|^2 \leq \sum_{n=1}^{\infty} |c_n|^2.$$

Hint: The closed graph theorem again, applied to the appropriate map

$$T : \ell^2 \to \mathcal{H}/N$$

where N is the orthogonal complement of the closed linear span of the vectors g_n.

3.37. A sequence of distinct vectors $\{h_n\}$ in a separable Hilbert space $\mathcal{H}$ is called a *frame* if there exist finite positive constants M_1 and M_2 with

$$M_1\|h\|^2 \leq \sum_{n=1}^{\infty} |\langle h, h_n \rangle|^2 \leq M_2\|h\|^2$$

for all $h \in \mathcal{H}$. Observe that if $\{h_n\}$ is a frame, then $\{h_n\}$ is a Bessel sequence (as defined in Exercise 3.36), and that whenever $\{h_n\}$ is a Bessel sequence, the second inequality in this definition must hold.

(a) Suppose that $\{h_n\}$ is a frame, and define T by

$$Th = \sum_{n=1}^{\infty} \langle h, h_n \rangle h_n.$$

Show that T is a bounded linear operator on $\mathcal{H}$.

(b) Show that
$$M_1\|h\|^2 \leq \langle Th, h \rangle \leq \|Th\| \cdot \|h\|$$

for all h, and thus that T is bounded below.

(c) Show that T is self-adjoint.

(d) Conclude from (b), (c), and Exercise 2.16 in Chapter 2 that T is invertible.

Frames are an important area of current research, and they have applications to signal processing, and image and data compression and analysis.

Chapter 4
Compact Operators

> The theory of compact operators is a convincing example that
> deep and important mathematics can be—or should I say must
> be—elegant.
> A. Pietsch ([34], p. 51).

To set the stage for the main topic of this chapter, we begin with a look at finite-dimensional spaces.

4.1 Finite-Dimensional Spaces

A vector space is finite-dimensional if it has a finite Hamel basis; that is, if it has a finite linearly independent spanning set. Finite-dimensional normed linear spaces—like $\mathbb{C}^n$ in your choice of norm—have some especially nice properties. On $\mathbb{C}^n$ we often use the norms

$$\|(z_1, z_2, \ldots, z_n)\|_2 = \left(\sum_{j=1}^{n} |z_j|^2 \right)^{1/2}$$

or

$$\|(z_1, z_2, \ldots, z_n)\|_1 = \sum_{j=1}^{n} |z_j|$$

or

$$\|(z_1, z_2, \ldots, z_n)\|_\infty = \max_{1 \le j \le n} |z_j|,$$

but as we will see these choices are all "equivalent" in a certain sense that we formally define below.

In $\mathbb{C}^n$, or any finite-dimensional normed linear space, the closed unit ball is compact. The statement in $\mathbb{C}^n$ may be known to the reader as the Heine–Borel theorem, and its extension to any finite-dimensional normed linear space will follow from Theorem 4.2 below. Recall that in a metric space a set A is compact in the open cover sense (every open cover has a finite subcover) if and only if it is limit point compact (every infinite subset of A has a limit point in A). The compactness of the closed unit ball fails in any infinite-dimensional normed linear space. For example, if $\mathscr{H}$ is a Hilbert space and $\{e_n\}_1^\infty$ is an infinite orthonormal set in $\mathscr{H}$, then since

B.D. MacCluer, *Elementary Functional Analysis*, DOI 10.1007/978-0-387-85529-5_4,   77
© Springer Science+Business Media, LLC 2009

$\|e_n - e_m\|^2 = 2$ whenever $n \neq m$, the set $\{e_n\}$ has no limit point. In Exercise 4.3 the analogous result in a normed linear space is outlined. As a consequence of this exercise, we obtain the conclusion that a normed linear space is finite-dimensional if and only if the closed unit ball is compact.

We introduced the notion of equivalent norms in the last chapter; here we make the formal definition.

Definition 4.1. Suppose X is a vector space and $\| \cdot \|_\beta$ and $\| \cdot \|_\gamma$ are two norms on X. We say that these norms are *equivalent* if there exist finite positive constants m and M with

$$m\|x\|_\beta \leq \|x\|_\gamma \leq M\|x\|_\beta$$

for all x in X.

Equivalence of norms in an equivalence relation and the topologies induced by two equivalent norms are the same. Thus topological concepts like compactness are unchanged when one norm is replaced by an equivalent one, and equivalent norms give rise to the same convergent sequences. We encourage the reader to verify these assertions.

Theorem 4.2. *In a finite-dimensional vector space, any two norms are equivalent.*

Proof. Let X be a finite-dimensional vector space and suppose $\| \cdot \|_\beta$ and $\| \cdot \|_\gamma$ are two norms on X. Fix a Hamel basis $\{b_1, b_2 \ldots, b_n\}$ for X and define a third norm on X as follows: Given $x \in X$ we write x uniquely in the form

$$x = \alpha_1 b_1 + \cdots + \alpha_n b_n$$

for scalars α_j and set

$$\|x\|_\infty = \max\{|\alpha_j| : 1 \leq j \leq n\}.$$

The reader can easily check that $\| \cdot \|_\infty$ is a norm on X. It suffices to show that both $\| \cdot \|_\beta$ and $\| \cdot \|_\gamma$ are equivalent to $\| \cdot \|_\infty$. We will verify that $\| \cdot \|_\beta$ and $\| \cdot \|_\infty$ are equivalent; the equivalence of $\| \cdot \|_\gamma$ and $\| \cdot \|_\infty$ will follow in exactly the same way.

For arbitrary $x = \alpha_1 b_1 + \cdots + \alpha_n b_n$ we have

$$\|x\|_\beta \leq \sum_1^n \|\alpha_j b_j\|_\beta = \sum_1^n |\alpha_j| \|b_j\|_\beta \leq \left(\sum_1^n \|b_j\|_\beta \right) \|x\|_\infty$$

so that $\|x\|_\beta \leq M\|x\|_\infty$ for $M = \sum \|b_j\|_\beta$. Now consider the unit sphere $S = \{x : \|x\|_\infty = 1\}$ in $(X, \| \cdot \|_\infty)$. Let

$$d = \inf\{\|x\|_\beta : x \in S\}.$$

We may find a sequence y_k of unit vectors in S with $\|y_k\|_\beta \to d$. Write each y_k as

$$y_k = \alpha_{1,k} b_1 + \cdots + \alpha_{n,k} b_n$$

and note that for all k and all $1 \leq j \leq n$, $|\alpha_{j,k}| \leq 1$ since y_k is in S. Since we are only concerned with finitely many j, we may find a subsequence $k_1, k_2, k_3, \ldots$ such that $\{a_{j,k_m}\}$ converges, as $m \to \infty$, for each $j = 1, 2, \ldots, n$. Denote the limit of $\{\alpha_{j,k_m}\}$ by α_j. The corresponding subsequence y_{k_m} of course still has $\|y_{k_m}\|_\beta \to d$. Set

$$y_0 = \sum_{j=1}^{n} \alpha_j b_j.$$

We claim that $y_{k_m} \to y_0$ in $\| \cdot \|_\infty$. This follows from the calculation

$$\|y_{k_m} - y_0\|_\infty = \|\sum_{j=1}^{n} (\alpha_{j,k_m} - \alpha_j) b_j\|_\infty = \max\{|\alpha_{j,k_m} - \alpha_j| : 1 \leq j \leq n\} \to 0.$$

This verifies the claim and shows that, in particular, $\|y_0\|_\infty = 1$ and thus $y_0 \neq 0$. Moreover, since by the first part of the proof

$$\|y_{k_m} - y_0\|_\beta \leq M \|y_{k_m} - y_0\|_\infty \to 0$$

we must have $\|y_0\|_\beta = \lim \|y_{k_m}\|_\beta = d$, so that $d \neq 0$. Finally for any nonzero x in X,

$$\frac{x}{\|x\|_\infty} \in S$$

and therefore

$$\left\| \frac{x}{\|x\|_\infty} \right\|_\beta \geq \|y_0\|_\beta$$

so that

$$\|x\|_\beta \geq \|y_0\|_\beta \|x\|_\infty = d\|x\|_\infty$$

for nonzero d, as desired. $\qquad\square$

Proposition 4.3. *Any finite-dimensional normed linear space is a Banach space, and any finite-dimensional subspace of a normed linear space is necessarily a closed subspace.*

Proof. Let $(X, \| \cdot \|)$ be the given normed linear space, and suppose that X is finite-dimensional. Fix a basis $\{b_1, b_2, \ldots, b_n\}$ and let $\| \cdot \|_\infty$ be a second norm defined on X as in the proof of Theorem 4.2. We leave it as Exercise 4.1 to check that $(X, \| \cdot \|)$ is complete if and only if $(X, \| \cdot \|_\infty)$ is complete. The first statement in the proposition will then follow if we can show that $(X, \| \cdot \|_\infty)$ is complete. Suppose that $\{y_m\}$ is a Cauchy sequence in $(X, \| \cdot \|_\infty)$ and for each m write

$$y_m = \sum_{j=1}^{n} \alpha_{j,m} b_j.$$

By the definition of $\| \cdot \|_\infty$, we must have that $\{\alpha_{j,m}\}_{m=1}^{\infty}$ is a Cauchy sequence of scalars for each j, $1 \leq j \leq n$. Hence there exists $\widetilde{\alpha}_j$ so that $\alpha_{j,m} \to \widetilde{\alpha}_j$ as $m \to \infty$, for each $1 \leq j \leq n$. Define $y_0 = \sum_{j=1}^{n} \widetilde{\alpha}_j b_j$. It is easy to see that

$$\|y_m - y_0\|_\infty \to 0$$

and hence $(X, \|\cdot\|_\infty)$ is complete.

Since the second statement of the proposition follows immediately from the first, this completes the proof.   □

Proposition 4.4. *Every linear map from a finite-dimensional normed linear space into a normed linear space is continuous.*

Proof. Suppose $T : X \to Y$ is as in the statement, and fix a basis $\{b_1, \ldots, b_n\}$ in X. Define a second norm $\|\cdot\|_\infty$ on X as in the proof of Theorem 4.2. The map T is continuous with respect to the original norm on X if and only if it is continuous with respect to the equivalent norm $\|\cdot\|_\infty$. We have

$$\|Tx\|_Y = \|T(\sum_{1}^{n} \alpha_k b_k)\|_Y \leq \sum_{1}^{n} |\alpha_k| \|Tb_k\|_Y$$

$$\leq \left(\max_{1 \leq k \leq n} |\alpha_k| \right) \left(\sum_{k=1}^{n} \|Tb_k\|_Y \right) = \left(\sum_{k=1}^{n} \|Tb_k\|_Y \right) \|x\|_\infty$$

establishing the boundedness, and hence the continuity, of T.   □

An alternate proof for Proposition 4.4 is outlined in Exercise 4.5.

4.2 Compact Operators

The idea motivating this section is to find a subspace of $\mathscr{B}(X,Y)$ consisting of operators which behave "like" linear maps on finite-dimensional spaces. One might naturally first think of singling out the operators that have finite-dimensional range. As we will see, this is not the most useful class of operators, so instead we make the following definition.

Definition 4.5. If X and Y are Banach spaces and $T : X \to Y$ is linear, we will say that T is *compact* if whenever $\{x_n\}$ is a bounded sequence in X, then $\{Tx_n\}$ has a convergent subsequence in Y.

Equivalently, T is compact if the image of any bounded set E in X under T has compact closure; the verification of this statement is left to the reader. The definition of compactness does not a priori require that the linear map T be bounded, but our first result will say this is so.

Proposition 4.6. *If T is compact, then T is bounded.*

Proof. If T is not bounded, we may find unit vectors v_n in X with $\|Tv_n\| \uparrow \infty$. This implies that $\{Tv_n\}$ cannot have a convergence subsequence, since if $Tv_{n_k} \to y$, then $\|Tv_{n_k}\| \to \|y\|$.   □

Example 4.7. The forward shift $S : \ell^2 \to \ell^2$ is not compact, since if e_n denotes the standard nth basis vector for ℓ^2, $\{Se_n\}$ has no convergent subsequence.

Example 4.8. Any linear operator $T : \mathbb{C}^n \to \mathbb{C}^n$ is compact. To see this, let $\{x_n\}$ be a bounded sequence of vectors in $\mathbb{C}^n$; say $\|x_n\| \leq M$. Since T is bounded by Proposition 4.4, $\{Tx_n\}$ is a set of vectors in the closed ball $\overline{B(0,R)}$ in $\mathbb{C}^n$, where $R = \|T\|M$. Since closed balls in $\mathbb{C}^n$ are compact, $\{Tx_n\}$ has a convergent subsequence.

The same idea can be used to show that if T is a bounded linear operator from X to Y, where X and Y are Banach spaces, and the range of T is a finite-dimensional subspace of Y, then T is compact. Such an operator T is called a *finite rank* operator.

The next result is easy, but important. Its proof is left as Exercise 4.6.

Proposition 4.9. *Let X be a Banach space, and suppose S is in $\mathcal{B}(X)$, and that T_1, T_2 are compact operators in $\mathcal{B}(X)$. The operators $T_1 + T_2$, ST_1, $T_1 S$, and αT_1 are compact, for any scalar α.*

This result says that the collection of all compact operators from X to X, which we will denote $\mathcal{K}(X)$, is a linear subspace in $\mathcal{B}(X)$ which is also a two-sided ideal (see Section 5.3 for more on this last terminology). That leads us to an important question: Is $\mathcal{K}(X)$ a *closed* subspace of $\mathcal{B}(X)$?

Theorem 4.10. *Suppose X is a Banach space. If $\{T_n\}$ is a sequence of compact operators in $\mathcal{B}(X)$ and $\|T_n - T\| \to 0$ for some $T \in \mathcal{B}(X)$, then T is compact.*

Proof. The argument we will use is sometimes referred to as the "diagonal trick," for reasons that should become apparent. Let $\{x_n\}$ be a bounded sequence in X. To show that T is compact, we must show that $\{Tx_n\}$ has a convergent subsequence. To do this, it suffices to show that $\{Tx_n\}$ has a subsequence which is Cauchy in X.

Since T_1 is compact, we may find a subsequence $\{x_{1,n}\}_{n=1}^{\infty}$ of $\{x_n\}$ such that $T_1(x_{1,n})$ converges in X as $n \to \infty$. Now $\{x_{1,n}\}_{n=1}^{\infty}$ is a bounded sequence and T_2 is compact, so there is a subsequence $\{x_{2,n}\}_{n=1}^{\infty}$ of $\{x_{1,n}\}_{n=1}^{\infty}$ such that $T_2(x_{2,n})$ converges. Of course we also have $T_1(x_{2,n})$ converging, since $\{x_{2,n}\}_{n=1}^{\infty}$ is a subsequence of $\{x_{1,n}\}_{n=1}^{\infty}$ and $T(x_{1,n})$ converges.

Continue, so that $\{x_{k,n}\}_{n=1}^{\infty}$ is a subsequence of $\{x_{k-1,n}\}_{n=1}^{\infty}$, with $T_k(x_{k,n})$ converging as $n \to \infty$, as well as $T_j(x_{k,n})$ for $j = 1, 2, \ldots, k-1$. Schematically we have

$$x_{1,1}\ x_{1,2}\ x_{1,3}\ \cdots$$

$$x_{2,1}\ x_{2,2}\ x_{2,3}\ \cdots$$

$$x_{3,1}\ x_{3,2}\ x_{3,3}\ \cdots$$

$$\vdots$$

$$x_{k,1}\ x_{k,2}\ x_{k,3}\ \cdots$$

where each row is a subsequence of the preceding rows, and when we apply the operators $T_1, T_2, \ldots, T_k$ to the kth row, a convergent sequence results.

Consider the diagonal sequence $\{x_{n,n}\}$, and note two things:

- This is a subsequence of the original sequence $\{x_n\}$.
- For each k, $\{T_k(x_{n,n})\}_{n=1}^{\infty}$ converges as $n \to \infty$, since $\{x_{n,n}\}_{n=k}^{\infty}$ is a subsequence of $\{x_{k,j}\}_{j=1}^{\infty}$.

This second property is just the observation that from the kth term on, $\{x_{n,n}\}$ is a subsequence of the kth row, and the operator T_k applied to the kth row produces a convergent subsequence.

We claim that $\{Tx_{n,n}\}_{n=1}^{\infty}$ converges. It is enough to show that it is a Cauchy sequence in X. We are given that $\|x_n\| \leq M$ for some finite value M. Let $\varepsilon > 0$ be given. Since $\|T - T_n\| \to 0$ we may find K so that $\|T - T_K\| < \varepsilon/(3M)$; K is now fixed by this requirement. Since $\{T_K x_{n,n}\}$ converges as $n \to \infty$, it is Cauchy and there exists N such that if $n, m \geq N$, then

$$\|T_K x_{n,n} - T_K x_{m,m}\| < \frac{\varepsilon}{3}. \tag{4.1}$$

Thus, for $n, m \geq N$,

$$\|Tx_{n,n} - Tx_{m,m}\| \leq \|Tx_{n,n} - T_K x_{n,n}\| + \|T_K x_{n,n} - T_K x_{m,m}\| + \|T_K x_{m,m} - Tx_{m,m}\|$$
$$\leq \frac{\varepsilon}{3M}M + \frac{\varepsilon}{3} + \frac{\varepsilon}{3M}M = \varepsilon,$$

as desired.   □

We look next at some examples. Suppose that A is a diagonal operator on a Hilbert space $\mathscr{H}$ with diagonal $\{\alpha_j\}$ such that $\alpha_j \to 0$; recall this means $Ae_j = \alpha_j e_j$ where $\{e_j\}$ is an orthonormal basis for $\mathscr{H}$. We claim that A is compact. To see this, let A_n be the diagonal operator with diagonal $\{\beta_j\}$ where $\beta_j = \alpha_j$ for $1 \leq j \leq n$ and $\beta_j = 0$ for $j > n$. Note that A_n is a finite rank operator, since the range of A_n is contained in the span of $\{e_1, e_2, \ldots, e_n\}$. Moreover, $A - A_n$ is a diagonal operator with diagonal $\{\gamma_n\}$ where $\gamma_j = 0$ if $1 \leq j \leq n$ and $\gamma_j = \alpha_j$ for $j > n$, so that $\|A - A_n\| = \sup_{j>n} |\alpha_j| \to 0$ as $n \to \infty$. This shows that A is the limit of finite rank operators, hence A is compact, by Theorem 4.10. This example also shows that the finite rank operators form a proper subclass of the compact operators.

It is also the case that a compact diagonal operator must have its diagonal converging to 0; see Exercise 4.8.

Although the result is true more generally, we will find it convenient in the next result to restrict our attention to Hilbert spaces with a countable orthonormal basis. Having a countable orthonormal basis is equivalent to being a *separable* Hilbert space, that is, having a countable dense subset; see Exercise 1.23 in Chapter 1. Nonseparable Hilbert spaces were not studied before 1934, and we will restrict our attention to the separable case whenever convenient.

Theorem 4.11. *If $T \in \mathcal{B}(\mathcal{H})$ is a compact operator on a Hilbert space $\mathcal{H}$ having a countable orthonormal basis, there exists finite rank operators T_n with $\|T - T_n\| \to 0$; that is, every compact operator is a limit of finite rank operators.*

Proof. There are three steps to the proof: Defining some likely candidates for the operators T_n, showing that for each fixed h in $\mathcal{H}$, $T_n h \to Th$, and then using this pointwise convergence to show $\|T_n - T\| \to 0$.

For the first step, suppose that $\{e_1, e_2, \ldots\}$ is an orthonormal basis for the closure of the range of T (a closed subspace of $\mathcal{H}$); notice it is only interesting if this basis is infinite. Let P_n be the projection onto the span of the first n vectors $e_1, e_2, \ldots, e_n$. Notice that we are using here the fact that this span is a closed subspace of $\mathcal{H}$, which follows from Proposition 4.3 or Exercise 1.21 in Chapter 1. Define $T_n = P_n T$, so that T_n is clearly a finite rank operator.

Now we check convergence of the T_n to T *at each point* of $\mathcal{H}$. Let h be any vector in $\mathcal{H}$ and set $k = Th$. We have

$$T_n h = P_n Th = P_n k = \sum_{j=1}^{n} \langle k, e_j \rangle e_j,$$

where we are using the result of Exercise 1.21 in Chapter 1, and

$$Th = \sum_{j=1}^{\infty} \langle k, e_j \rangle e_j,$$

since $\{e_j\}$ is an orthonormal basis for the closure of the range of T. Thus

$$\|T_n h - Th\|^2 = \sum_{j=n+1}^{\infty} |\langle k, e_j \rangle|^2,$$

which tends to 0 as $n \to \infty$, since $\sum_{j=1}^{\infty} |\langle k, e_j \rangle|^2 = \|Th\|^2 < \infty$.

Having established this pointwise convergence of T_n to T, we now consider $\|T_n - T\|$. Let B_c denote the closed unit ball in $\mathcal{H}$. Since T is a compact operator, the closure of $T(B_c)$ is compact. Given $\varepsilon > 0$ the collection of open balls $B(Th, \varepsilon)$, centered at points Th for $h \in B_c$ and with radius ε, forms an open cover of the closure of $T(B_c)$. By compactness, we may select a finite subcover, say

$$\overline{T(B_c)} \subseteq \bigcup_{j=1}^{m} B(Th_j, \varepsilon), \tag{4.2}$$

for some positive integer m and some $h_j \in B_c$. By our pointwise estimate, for each j, $1 \le j \le m$, there is an integer $N(j)$ so that

$$\|T_n h_j - Th_j\| < \varepsilon$$

if $n \ge N(j)$. Set $N_* = \max_{1 \le j \le m} N(j)$ and let h be an arbitrary unit vector in $\mathcal{H}$. By condition (4.2) we may find a value of j, $1 \le j \le m$, so that $\|Th - Th_j\| < \varepsilon$. For any $n \ge N_*$ we have

$$\|T_n h - Th\| \leq \|T_n h - T_n h_j\| + \|T_n h_j - Th_j\| + \|Th_j - Th\|$$
$$= \|P_n T(h - h_j)\| + \|T_n h_j - Th_j\| + \|T(h - h_j)\|$$
$$\leq 2\|T(h - h_j)\| + \varepsilon$$
$$\leq 3\varepsilon,$$

where we have used the fact that the projection P_n has norm 1. Since h was an arbitrary unit vector, this calculation shows that if $n \geq N_*$, then $\|T_n - T\| \leq 3\varepsilon$, and since ε is arbitrary, this shows that $\|T_n - T\| \to 0$ as $n \to \infty$, as desired. □

This theorem is true even without the hypothesis that $\mathcal{H}$ has a countable basis, that is, even for nonseparable Hilbert spaces; the necessary additions to the proof for this case can be found in [8]. It makes sense to ask what happens if $\mathcal{H}$ is replaced by a Banach space X, i.e., is every compact operator in $\mathcal{B}(X)$ a limit of finite rank operators? This question, known as the approximation problem, was formulated by T. Hildebrandt in 1931, and has been a problem of fundamental importance in Banach space theory. Over time, many equivalent properties were discovered, and various related approximation properties were defined. In 1973, Per Enflo caused a sensation by constructing a counterexample to the original approximation problem. Enflo's work on this problem also yielded a negative answer to another long-open problem in functional analysis: Does every separable Banach space have a Schauder basis, that is, in any Banach space X is there always a sequence $\{x_j\}$ such that each $x \in X$ can be uniquely written as

$$x = \sum_{j=1}^{\infty} c_j x_j = \lim_{n \to \infty} \left(\sum_{j=1}^{n} c_j x_j \right)?$$

It *is* true that in a Banach space with a Schauder basis, every compact operator is a limit of finite rank operators, and this includes all of the familiar Banach spaces.

Enflo's work provided a solution to Problem #153 in the "Scottish Problem Book", posed by S. Mazur in 1936, for which a prize of "a live goose" had been offered[1]. When Enflo was lecturing in Warsaw in 1972, Mazur presented him with the prize goose promised 36 years earlier.

The next result is sometimes phrased as "$\mathcal{K}(\mathcal{H})$ is self-adjoint" when $\mathcal{H}$ is a Hilbert space.

Proposition 4.12. *If T is in $\mathcal{B}(\mathcal{H})$ for a separable Hilbert space $\mathcal{H}$, then T is compact if and only if T^* is compact.*

Proof. Since $T^{**} = T$, it suffices to prove that T compact implies T^* is compact. By Theorem 4.11, if T is compact, there are finite rank operators T_n that converge to T. Now $\|T_n - T\| = \|T_n^* - T^*\|$ by Proposition 2.14, and we will be done by an appeal to Theorem 4.10 if we can show that each T_n^* is finite rank. Let P_n be the projection of $\mathcal{H}$ onto the range of T_n, a closed subspace of $\mathcal{H}$. Each P_n is finite rank since each T_n is. Since P_n is a projection, $P_n T_n = T_n$, and taking adjoints, $T_n^* P_n^* = T_n^*$. But

---

[1] See Section 3.5.

projections are self-adjoint, so we have $T_n^* = T_n^* P_n$. It is easy to see that the finite rank operators form a (two-sided) ideal (see Exercise 4.6), so we conclude that T_n^* is finite rank, and thus T^* is compact. □

Since Theorem 4.11 extends to nonseparable Hilbert spaces, so does Proposition 4.12. The implication "T compact implies T^* compact" also extends to the Banach space setting, where it sometimes goes by the name of "Schauder's Theorem"; see [8].

To explore a large class of compact operators, we give a definition.

Definition 4.13. Suppose that $\mathcal{H}$ is a Hilbert space with a countable orthonormal basis, and let T be in $\mathcal{B}(\mathcal{H})$. We say that T is *Hilbert–Schmidt* if there is an orthonormal basis $\{e_n\}_1^\infty$ of $\mathcal{H}$ such that $\sum_{n=1}^\infty \|Te_n\|^2 < \infty$.

It is convenient to know that the sum appearing in this definition is actually independent of the choice of basis; this is the next result, which is due to von Neumann.

Proposition 4.14. *Suppose that T is a bounded linear operator on a separable Hilbert space $\mathcal{H}$ and $\{e_n\}_{n=1}^\infty$ is an orthonormal basis for $\mathcal{H}$ such that*

$$\sum_{n=1}^\infty \|Te_n\|^2 < \infty.$$

For any other orthonormal basis $\{f_n\}_{n=1}^\infty$, we have

$$\sum_{n=1}^\infty \|Tf_n\|^2 = \sum_{n=1}^\infty \|Te_n\|^2.$$

Proof. The proof relies on repeated applications of Parseval's identity. For each n we have

$$Tf_n = \sum_{j=1}^\infty \langle Tf_n, f_j\rangle f_j \text{ and } \|Tf_n\|^2 = \sum_{j=1}^\infty |\langle Tf_n, f_j\rangle|^2. \tag{4.3}$$

Similarly, for each n and each j,

$$\|Te_n\|^2 = \sum_{j=1}^\infty |\langle Te_n, f_j\rangle|^2, \tag{4.4}$$

$$\|T^*f_j\|^2 = \sum_{n=1}^\infty |\langle T^*f_j, e_n\rangle|^2, \tag{4.5}$$

$$\|T^*f_j\|^2 = \sum_{n=1}^\infty |\langle T^*f_j, f_n\rangle|^2. \tag{4.6}$$

Thus

$$\sum_{n=1}^\infty \|Tf_n\|^2 = \sum_{n=1}^\infty \sum_{j=1}^\infty |\langle Tf_n, f_j\rangle|^2 = \sum_{j=1}^\infty \sum_{n=1}^\infty |\langle T^*f_j, f_n\rangle|^2 = \sum_{j=1}^\infty \|T^*f_j\|^2,$$

where we have used Equations (4.3) and (4.6). Similarly, by Equations (4.4) and (4.5)

$$\sum_{n=1}^{\infty} \|Te_n\|^2 = \sum_{n=1}^{\infty} \sum_{j=1}^{\infty} |\langle Te_n, f_j \rangle|^2 = \sum_{j=1}^{\infty} \sum_{n=1}^{\infty} |\langle T^* f_j, e_n \rangle|^2 = \sum_{j=1}^{\infty} \|T^* f_j\|^2$$

so that

$$\sum_{n=1}^{\infty} \|T f_n\|^2 = \sum_{n=1}^{\infty} \|Te_n\|^2,$$

as desired. □

As a corollary of the proof of the last result, we see that if T is Hilbert–Schmidt, then so is T^*. The next result shows that a Hilbert–Schmidt operator is compact.

Theorem 4.15. *Every Hilbert–Schmidt operator on a separable Hilbert space is compact.*

Proof. Let A be Hilbert–Schmidt on $\mathcal{H}$. We will exhibit A as a limit of finite rank operators and use Theorem 4.10 to conclude A is compact. Fix an orthonormal basis $\{e_k\}_1^{\infty}$ of $\mathcal{H}$, so that $\sum_{k=1}^{\infty} \|Ae_k\|^2 < \infty$. For each $n \geq 1$, define A_n by

$$A_n h = A_n \left(\sum_{k=1}^{\infty} \hat{h}(k) e_k \right) = \sum_{k=1}^{n} \hat{h}(k) Ae_k,$$

where $\hat{h}(k) = \langle h, e_k \rangle$. Clearly A_n is linear, and A_n is finite rank, since the range of A_n is contained in the span of the vectors $Ae_1, Ae_2, \ldots, Ae_n$. Moreover, for any $h \in \mathcal{H}$,

$$\|(A - A_n)h\| = \| \sum_{k=n+1}^{\infty} \hat{h}(k) Ae_k \| \leq \left(\sum_{k=n+1}^{\infty} |\hat{h}(k)|^2 \right)^{\frac{1}{2}} \left(\sum_{k=n+1}^{\infty} \|Ae_k\|^2 \right)^{\frac{1}{2}}$$

so that

$$\|A - A_n\| \leq \left(\sum_{k=n+1}^{\infty} \|Ae_k\|^2 \right)^{\frac{1}{2}},$$

which tends to 0 as $n \to \infty$. □

The next result gives a large class of examples of Hilbert–Schmidt operators, and explains how they arise naturally. It concerns operators on $L^2(X, \mu)$ for some measure space (X, μ).

Theorem 4.16. *Suppose that $L^2(X, \mu)$ is a separable Hilbert space and K is an integral operator on $L^2(X, \mu)$, with kernel $k(x, y) \in L^2(X \times X)$. The operator K is Hilbert–Schmidt.*

Proof. Let $\{e_n\}_{n=1}^{\infty}$ be a basis for $L^2(X, \mu)$. For fixed $x \in X$, write $k_x(y) = k(x, y)$; then k_x is in $L^2(\mu)$ for almost every x and we have

$$(Ke_n)(x) = \int_X k(x,y)e_n(y)d\mu(y)$$

$$= \int_X k_x(y)e_n(y)d\mu(y)$$

$$= \langle k_x, \overline{e_n} \rangle,$$

so that

$$\|Ke_n\|^2 = \int_X |Ke_n(x)|^2 d\mu(x) = \int_X |\langle k_x, \overline{e_n} \rangle|^2 d\mu(x)$$

and

$$\sum_{n=1}^{\infty} \|Ke_n\|^2 = \sum_{n=1}^{\infty} \int_X |\langle k_x, \overline{e_n} \rangle|^2 d\mu(x) = \int_X \sum_{n=1}^{\infty} |\langle k_x, \overline{e_n} \rangle|^2 d\mu(x).$$

Now it is easy to see that $\{\overline{e_n}\}_{n=1}^{\infty}$ is also an orthonormal basis for $L^2(X, \mu)$, so that $\langle k_x, \overline{e_n} \rangle$ are the Fourier coefficients of k_x with respect to this basis. By Parseval's identity,

$$\sum_{n=1}^{\infty} |\langle k_x, \overline{e_n} \rangle|^2 = \|k_x\|^2.$$

Thus we have

$$\sum_{n=1}^{\infty} \|Ke_n\|^2 = \int_X \|k_x\|^2 d\mu(x)$$

$$= \int_X \left(\int_X |k(x,y)|^2 d\mu(y) \right) d\mu(x)$$

$$= \int_{X \times X} |k(x,y)|^2 d(\mu \times \mu) < \infty$$

since $k \in L^2(\mu \times \mu)$.   □

4.3 A Preliminary Spectral Theorem

Recall from linear algebra, that if M is a self-adjoint $n \times n$ matrix (meaning it is equal to its conjugate transpose M^*) then all the eigenvalues of M are real, and there is a unitary matrix U (meaning $U^* = U^{-1}$) such that UMU^{-1} is real and diagonal. There is an orthonormal basis for $\mathbb{C}^n$ consisting of eigenvectors of M. In fact, if M is normal, that is, if $M^*M = MM^*$, then M is unitarily diagonalizable and $\mathbb{C}^n$ has an orthonormal basis of eigenvectors. Results that are generalizations of this to operators on Hilbert spaces are called "spectral theorems." They appear with several quite distinct-looking formulations, and their connection to the finite-dimensional linear algebra results can seem somewhat obscured.

In this section we will obtain a spectral theorem for *compact self-adjoint operators* on a Hilbert space; one can view this as a complete description of such operators. Later, in Chapter 6, we will obtain a much more general spectral theorem for bounded normal operators, at which point we will revisit the main results of this section.

Our first goal is to obtain some information about the eigenvalues of compact self-adjoint operators in $\mathscr{B}(\mathscr{H})$. The notion of an eigenvalue of an operator in $\mathscr{B}(\mathscr{H})$ is the expected one, and the definition looks the same for operators on Hilbert or Banach spaces.

Definition 4.17. We say that $\lambda \in \mathbb{C}$ is an *eigenvalue* of $T \in \mathscr{B}(X)$, where X is a Banach space, if there is a nonzero vector x in X so that $Tx = \lambda x$.

When λ is an eigenvalue of $T \in \mathscr{B}(X)$, the kernel of $T - \lambda I$ is called the *eigenspace* corresponding to the eigenvalue λ, and the nonzero vectors in the eigenspace are called eigenvectors. For a compact operator T with nonzero eigenvalue λ, the kernel of $T - \lambda I$ is necessarily finite-dimensional; see Exercise 4.10 and Exercise 4.25.

We will show that a compact self-adjoint operator T always has either $\|T\|$ or $-\|T\|$ as an eigenvalue. Before proceeding to the proof of this, we need a lemma. The role of self-adjointness in it, and in the next several results, is contained in the following observation: If T is self-adjoint, then for any vectors x and y, $\langle Tx, y \rangle = \langle x, Ty \rangle = \overline{\langle Ty, x \rangle}$, and so in particular $\langle Tz, z \rangle$ must be real for all vectors z.

For any operator T in $\mathscr{B}(\mathscr{H})$, it is easy to see that

$$\|T\| = \sup\{|\langle Tx, y \rangle| : \|x\| = 1, \|y\| = 1\}.$$

The next result refines this for self-adjoint operators.

Lemma 4.18. *Suppose T is a self-adjoint operator in $\mathscr{B}(\mathscr{H})$ for some Hilbert space $\mathscr{H}$. We have*

$$\|T\| = \sup_{\|x\|=1} |\langle Tx, x \rangle|.$$

Proof. Set $M = \sup_{\|x\|=1} |\langle Tx, x \rangle|$. Our goal is to show $M = \|T\|$. We make three easy observations

(a)   For each $h \neq 0 \in \mathscr{H}$, $|\langle Th, h \rangle| = |\langle T(\|h\| \frac{h}{\|h\|}), \|h\| \frac{h}{\|h\|} \rangle| \leq M \|h\|^2$.

(b)   Since $|\langle Th, h \rangle| \leq \|Th\| \|h\| \leq \|T\| \|h\|^2$ we must have $M \leq \|T\|$, by the definition of M.

(c)   For all $f, g \in \mathscr{H}$,

$$\langle T(f+g), f+g \rangle - \langle T(f-g), f-g \rangle = 4\mathrm{Re}\, \langle Tf, g \rangle.$$

This follows from expanding $\langle T(f \pm g), f \pm g \rangle$, using the self-adjointness of T to write

$$\langle Tf, g \rangle + \langle Tg, f \rangle = \langle Tf, g \rangle + \overline{\langle Tf, g \rangle} = 2\mathrm{Re}\, \langle Tf, g \rangle.$$

Since T is self-adjoint, $\langle Tx, x \rangle$ and $\langle Ty, y \rangle$ are real for any $x, y \in \mathscr{H}$, and we have

$$\langle Tx, x \rangle - \langle Ty, y \rangle \leq |\langle Tx, x \rangle| + |\langle Ty, y \rangle| \tag{4.7}$$
$$\leq M(\|x\|^2 + \|y\|^2)$$
$$= \frac{M}{2}(\|x - y\|^2 + \|x + y\|^2),$$

where we have used the observation in (a) and the parallelogram equality. Now let v be any unit vector and suppose $Tv \neq 0$. Let $s = \|Tv\|$ and put

$$x = v + \frac{1}{s}Tv \text{ and } y = v - \frac{1}{s}Tv$$

in Equation (4.7). We obtain

$$\langle Tx, x \rangle - \langle Ty, y \rangle \leq \frac{M}{2}\left(\left\|\frac{2}{s}Tv\right\|^2 + \|2v\|^2\right) = 4M.$$

Since by the calculation in (c) we have

$$\langle Tx, x \rangle - \langle Ty, y \rangle = 4\mathrm{Re}\left\langle Tv, \frac{1}{s}Tv \right\rangle = 4\|Tv\|$$

we conclude that $\|Tv\| \leq M$ for any unit vector v and hence $\|T\| \leq M$. Since we had already observed the reverse inequality, we are done. $\qquad\square$

Theorem 4.19. *If T is a compact self-adjoint operator in $\mathcal{B}(\mathcal{H})$, then at least one of the numbers $\|T\|$ and $-\|T\|$ is an eigenvalue of T.*

Proof. Without loss of generality we assume $\|T\| \neq 0$, else $T = 0$ and 0 is trivially an eigenvalue of T. By Lemma 4.18 we have $\|T\| = \sup_{\|x\|=1} |\langle Tx, x \rangle|$. Find unit vectors x_n with $|\langle Tx_n, x_n \rangle| \to \|T\|$. Since T is self-adjoint, each $\langle Tx_n, x_n \rangle$ is real, and passing to a subsequence if necessary (which we don't relabel) we may assume that $\langle Tx_n, x_n \rangle \to \lambda$ where either $\lambda = \|T\|$ or $\lambda = -\|T\|$. Since λ is real, we have

$$\|Tx_n - \lambda x_n\|^2 = \|Tx_n\|^2 - 2\mathrm{Re}\,\langle Tx_n, \lambda x_n \rangle + |\lambda|^2\|x_n\|^2$$
$$= \|Tx_n\|^2 - 2\lambda\langle Tx_n, x_n \rangle + \lambda^2$$
$$\leq \|T\|^2 - 2\lambda\langle Tx_n, x_n \rangle + \lambda^2$$
$$= 2\lambda^2 - 2\lambda\langle Tx_n, x_n \rangle.$$

As $n \to \infty$, $2\lambda^2 - 2\lambda\langle Tx_n, x_n \rangle \to 0$, so that

$$Tx_n - \lambda x_n \to 0. \tag{4.8}$$

Since T is compact, $\{Tx_n\}$ has a convergent subsequence, say $Tx_{n_k} \to y$. By (4.8), we have $\lambda x_{n_k} \to y$, and thus $\lambda Tx_{n_k} \to Ty$. But $\lambda Tx_{n_k} \to \lambda y$, so that $Ty = \lambda y$. Are we done? Yes, if we can show that $y \neq 0$. We have

$$\|Tx_{n_k}\| = \|\lambda x_{n_k} + Tx_{n_k} - \lambda x_{n_k}\|$$

$$\geq \|\lambda x_{n_k}\| - \|T x_{n_k} - \lambda x_{n_k}\|$$
$$= |\lambda| - \|T x_{n_k} - \lambda x_{n_k}\|,$$

where we know that $\|T x_{n_k} - \lambda x_{n_k}\| \to 0$ and $\lambda \neq 0$. But $T x_{n_k} \to y$, so that $\|y\| = \lim_{n \to \infty} \|T x_{n_k}\| \neq 0$. $\square$

The next result applies to any self-adjoint operator on a Hilbert space, compact or not.

Theorem 4.20. *Suppose that T is self-adjoint in $\mathscr{B}(\mathscr{H})$. Every eigenvalue of T is real, and the eigenvectors for distinct eigenvalues are orthogonal.*

Proof. Suppose that for some nonzero vector h and some scalar λ, $Th = \lambda h$. Since $\langle Th, h \rangle = \langle \lambda h, h \rangle = \lambda \|h\|^2$ and $\langle Th, h \rangle = \langle h, Th \rangle = \langle h, \lambda h \rangle = \overline{\lambda} \|h\|^2$ we must have $\lambda = \overline{\lambda}$, and λ is real.

If λ and μ are distinct (real) eigenvalues for T with $Th = \lambda h$ and $Tg = \mu g$ then $0 = \langle Th, g \rangle - \langle h, Tg \rangle = \lambda \langle h, g \rangle - \mu \langle h, g \rangle = (\lambda - \mu) \langle h, g \rangle$ and $\langle h, g \rangle = 0$. $\square$

The reader is cautioned that this result does not say that a self-adjoint operator must *have* eigenvalues. In contrast to the conclusion of Theorem 4.19, neither compactness of T nor self-adjointness of T is (separately) sufficient to guarantee the existence of an eigenvalue for T. Exercises 4.14 and 4.15 ask you to verify this assertion.

Theorem 4.21. *Suppose that T is a compact self-adjoint operator in $\mathscr{B}(\mathscr{H})$. The set of eigenvalues of T is a finite or countably infinite set of real numbers; if infinite, the eigenvalues form a sequence that converges to zero.*

Proof. By Theorem 4.20, all the eigenvalues are real. Also observe that if $Tx = \lambda x$, then $|\lambda| \leq \|T\|$, so that no eigenvalue has absolute value greater than $\|T\|$. There is nothing further to do if the set of eigenvalues is finite, so suppose it is infinite. We claim that for each $\varepsilon > 0$, there are at most finitely many eigenvalues with absolute value at least ε. Suppose this is not the case. We may then find a sequence $\{\lambda_j\}$ of distinct eigenvalues, with $|\lambda_j| \geq \varepsilon$ and *unit* eigenvectors y_j with $Ty_j = \lambda_j y_j$. By Theorem 4.20 the y_j are orthogonal, and thus

$$\|Ty_j - Ty_k\|^2 = \|\lambda_j y_j - \lambda_k y_k\|^2 = |\lambda_j|^2 + |\lambda_k|^2 \geq 2\varepsilon^2.$$

This is a contradiction, since the compactness of T guarantees that $\{Ty_j\}$ has a convergent subsequence. Thus the claim is verified and we have shown:

(a)   The set of eigenvalues is countable, since $\{\lambda : \lambda$ is an eigenvalue and $|\lambda| \geq 1/n\}$ is finite for every positive integer n.

(b)   If the eigenvalues are a countably infinite set, they form a sequence which converges to zero.

This completes the proof. $\square$

A version of this result holds for arbitrary compact operators on a Hilbert or Banach space. The eigenvalues need not be real, but if infinite they still form a sequence converging to zero.

In the next result, we write TM for $\{Tm : m \in M\}$.

Lemma 4.22. *Suppose that T is a bounded operator on a Hilbert space $\mathcal{H}$ and that M is a closed subspace of $\mathcal{H}$. If $TM \subseteq M$, then $T^*M^\perp \subseteq M^\perp$. Conversely, if $T^*M^\perp \subseteq M^\perp$, then $TM \subseteq M$.*

Proof. Since $T^{**} = T$ and $(M^\perp)^\perp = M$, only the first assertion needs to be verified. Let n be in $M^\perp$ and let m be in M. We must show that $T^*n \perp m$, or equivalently, $\langle T^*n, m \rangle = 0$. We have

$$\langle T^*n, m \rangle = \langle n, Tm \rangle = 0$$

since Tm is in M.   □

The next corollary is immediate.

Corollary 4.23. *If T is a self-adjoint operator in $\mathcal{B}(\mathcal{H})$, and if $TM \subseteq M$ for some closed subspace M, then $TM^\perp \subseteq M^\perp$.*

A closed subspace M is called an *invariant subspace* for $T \in \mathcal{B}(\mathcal{H})$ if $TM \subseteq M$. It is called a *reducing subspace* for T if both $TM \subseteq M$ and $TM^\perp \subseteq M^\perp$. By virtue of Lemma 4.22, M is reducing if and only if it is invariant for both T and T^*. For an easy example, note that the subspaces

$$N_E = \{f \in L^2[0,1] : f(x) = 0 \text{ almost everywhere on } E\}$$

for any measurable subset E of $[0,1]$ are reducing subspaces for the operator M_x of multiplication by $\varphi(x) = x$ on $(L^2[0,1], dx)$. On the other hand,

$$N = \{f \in L^2_a(\mathbb{D}) : f(0) = 0\}$$

is an invariant subspace for the multiplication operator M_z on $L^2_a(\mathbb{D})$, but it is not a reducing subspace since, for example, $M_z^*(z)$ is the constant $1/2$, which is not in N. Further examples can be found in Exercise 4.17.

The terminology "reducing subspace" is suggestive of how these subspaces are used: Since $\mathcal{H} = M \oplus M^\perp$, if M is a reducing subspace of T then the study of T on $\mathcal{H}$ is "reduced" to its study on the (smaller) Hilbert spaces M and $M^\perp$. We'll see this in action in the next result, which is the spectral theorem for compact self-adjoint operators. Our presentation follows that in [48].

Theorem 4.24 (Spectral Theorem, Preliminary Version). *Let $T \neq 0$ be a compact, self-adjoint operator in $\mathcal{B}(\mathcal{H})$. There exists a finite or countably infinite orthonormal set $\{g_n\}$ of eigenvectors of T, with corresponding real eigenvalues $\{\lambda_n\}$, such that*

$$Tx = \sum_n \lambda_n \langle x, g_n \rangle g_n.$$

If $\{\lambda_n\}$ is infinite, then $\lambda_n \to 0$.

Proof. The fact that all eigenvalues are real has already been shown. For the rest we give an inductive construction. We know that $\lambda_1 \equiv \|T\|$ or $-\|T\|$ is an eigenvalue of T. Pick a corresponding unit eigenvector g_1. Let M_1 be the span of $\{g_1\}$. Since $T(\alpha g_1) = \alpha \lambda_1 g_1$, M_1 is an invariant subspace for T. By Corollary 4.23, it is a reducing subspace, i.e., $TM_1^\perp \subseteq M_1^\perp$. Let T_2 be the restriction of T to the Hilbert space $M_1^\perp \equiv \mathscr{H}_2$. Now, the restriction of a compact operator to an invariant subspace is compact (see Exercise 4.21), so $T_2 \in \mathscr{B}(\mathscr{H}_2)$ is compact. We claim that T_2 is also self-adjoint: if x, y are in $M_1^\perp$, then

$$\langle T_2^* x, y \rangle = \langle x, T_2 y \rangle = \langle x, Ty \rangle = \langle Tx, y \rangle = \langle T_2 x, y \rangle,$$

where the fact that T is self-adjoint is used. Applying Theorem 4.19 again, this time to $T_2 \in \mathscr{B}(\mathscr{H}_2)$, we see that T_2 has an eigenvalue λ_2 with $\lambda_2 = \|T_2\|$ or $-\|T_2\|$, and corresponding unit eigenvector $g_2 \in \mathscr{H}_2$. Notice that $|\lambda_2| \leq |\lambda_1|$, and of course, λ_2 is also an eigenvalue of the original operator T. Since g_2 is in $\mathscr{H}_2 = M_1^\perp$, $g_2 \perp g_1$.

Proceed inductively: Suppose we have obtained pairwise orthogonal unit eigenvectors $g_1, g_2, \ldots, g_n$ of T corresponding to real eigenvalues $\lambda_1, \lambda_2, \ldots, \lambda_n$ with $|\lambda_j| = \|T_j\|$, where $T_1 = T$, and T_j is the restriction of T to

$$[\operatorname{span} \{g_1, g_2, \ldots, g_{j-1}\}]^\perp$$

for $1 < j \leq n$. Let T_{n+1} be the restriction of T to

$$\mathscr{H}_{n+1} \equiv [\operatorname{span} \{g_1, g_2, \ldots, g_n\}]^\perp.$$

Since span $\{g_1, g_2, \ldots, g_n\}$ is invariant under T, so is $\mathscr{H}_{n+1}$, by Corollary 4.23. Moreover, as above, $T_{n+1} : \mathscr{H}_{n+1} \to \mathscr{H}_{n+1}$ is compact and self-adjoint, and thus T_{n+1} must have eigenvalue λ_{n+1}, equal to either $\|T_{n+1}\|$ or $-\|T_{n+1}\|$, and we choose a corresponding unit eigenvector $g_{n+1} \in \mathscr{H}_{n+1}$; this is of course also an eigenvector for T on $\mathscr{H}$. By the definition of $\mathscr{H}_{n+1}$, g_{n+1} is orthogonal to g_j for $1 \leq j \leq n$.

As we continue this process, one of two things will happen. Either there is a smallest m with $T_m = 0$, in which case the process terminates with the construction of g_{m-1}, or $T_n \neq 0$ for all n. In the first case, consider, for arbitrary x in $\mathscr{H}$,

$$y \equiv x - \sum_{j=1}^{m-1} \langle x, g_j \rangle g_j.$$

Since $\sum_1^{m-1} \langle x, g_j \rangle g_j$ is the projection of x onto span $\{g_1, g_2, \ldots, g_{m-1}\}$, the projection theorem says that y is in the orthogonal complement of this span, that is, y is in $\mathscr{H}_m$. Thus we have

$$0 = T_m y = Ty = Tx - \sum_{j=1}^{m-1} \langle x, g_j \rangle T g_j = Tx - \sum_{j=1}^{m-1} \lambda_j \langle x, g_j \rangle g_j,$$

which says that

$$Tx = \sum_{j=1}^{m-1} \lambda_j \langle x, g_j \rangle g_j$$

for every x in $\mathcal{H}$, giving the desired conclusion in this case.

In the case that T_n is not zero for any n, note while there may be repeated values in the sequence $\lambda_1, \lambda_2, \ldots$, by Exercise 4.10 each value appears only finitely many times. This observation, together with Theorem 4.21, says $\lambda_n \to 0$. Again consider an arbitrary x in $\mathcal{H}$. We wish to show

$$Tx = \sum_{j=1}^{\infty} \lambda_j \langle x, g_j \rangle g_j,$$

i.e., that

$$Tx = \lim_{n \to \infty} \sum_{j=1}^{n-1} \lambda_j \langle x, g_j \rangle g_j.$$

To this end, set

$$y_n = x - \sum_{j=1}^{n-1} \langle x, g_j \rangle g_j$$

and notice that y_n is in $\mathcal{H}_n$ and $\sum_1^{n-1} \langle x, g_j \rangle g_j$ is in its orthogonal complement. In particular, this guarantees by the Pythagorean theorem that $\|x\| \geq \|y_n\|$. Now

$$\|Ty_n\| = \|T_n y_n\| \leq \|T_n\| \|y_n\| = |\lambda_n| \|y_n\| \leq |\lambda_n| \|x\|.$$

We know that $|\lambda_n| \to 0$ as $n \to \infty$ so we must have

$$\left\| Tx - \sum_{j=1}^{n-1} \lambda_j \langle x, g_j \rangle g_j \right\| = \|Ty_n\| \leq |\lambda_n| \|x\| \to 0,$$

which is our desired conclusion. $\qquad\square$

We note that the λ_n appearing in the statement of the last result must form a complete list of all the nonzero eigenvalues of T. To see this, observe that if $Tz = \mu z$ for some μ distinct from all the λ_n, then $z \perp g_n$ for all n, since the eigenvectors corresponding to distinct eigenvalues are necessarily orthogonal for any self-adjoint operator (Theorem 4.20). Hence

$$\sum \lambda_n \langle z, g_n \rangle g_n = 0 = Tz = \mu z$$

and $z = 0$.

In the last result, the g_n are an orthonormal sequence, but need not be an orthonormal basis for $\mathcal{H}$. The next result explores this further.

Corollary 4.25. *If T is a compact self-adjoint operator on a separable Hilbert space $\mathcal{H}$, then there is an orthonormal basis $\{e_n\}$ of $\mathcal{H}$ consisting of eigenvectors for T such that*

$$Tx = \sum_n \lambda_n \langle x, e_n \rangle e_n$$

for every x in $\mathscr{H}$, where λ_n is the eigenvalue of T corresponding to the eigenvector e_n. This sum is either finite or countably infinite.

Proof. By Theorem 4.24, there is a finite or infinite orthonormal sequence $\{g_n\}$ such that

$$Tx = \sum_n \lambda_n \langle x, g_n \rangle g_n \tag{4.9}$$

and $Tg_n = \lambda_n g_n$. By the construction in Theorem 4.24, the λ_n are nonzero. Let $\{h_m\}$ be an orthonormal basis for ker T; this is at most countable since $\mathscr{H}$ is assumed to be separable. We have $Th_m = 0$ and each h_m is an eigenvector of T. Since eigenvectors corresponding to distinct eigenvalues are orthogonal, $h_m \perp g_n$ for all m, n. Thus $\{g_n\} \cup \{h_m\}$ is an orthonormal set in $\mathscr{H}$ consisting of eigenvectors of T. We claim it is an orthonormal basis for $\mathscr{H}$. By Equation (4.9),

$$x - \sum_n \langle x, g_n \rangle g_n$$

is in ker T so that

$$x - \sum_n \langle x, g_n \rangle g_n = \sum_m c_m h_m$$

for some coefficients c_m; in fact we must have

$$c_m = \langle x - \sum_n \langle x, g_n \rangle g_n, h_m \rangle = \langle x, h_m \rangle$$

since $h_m \perp g_n$. We have shown that an arbitrary $x \in \mathscr{H}$ can be written as

$$x = \sum_n \langle x, g_n \rangle g_n + \sum_m \langle x, h_m \rangle h_m.$$

Thus $\{g_n\} \cup \{h_m\}$ is a countable orthonormal basis for $\mathscr{H}$. If we relabel this as $\{e_n\}$, we are done. $\square$

4.4 The Invariant Subspace Problem

The invariant subspace problem, which has been variously described as "the most fundamental question in operator theory" [3] or "the most famous unsolved problem in the theory of bounded linear operators" [35], asks whether every $T \in \mathscr{B}(X)$ has a nontrivial closed invariant subspace; in this section the term "invariant subspace" will always mean a nontrivial closed invariant subspace. One can ask this question when X is a Banach space or when X is a Hilbert space and this distinction is important. Of course, if T has an eigenvalue, then the corresponding eigenspace is invariant for T, but it is easy to give examples of operators with no eigenvalues but yet having invariant subspaces (see Exercises 4.14 and 4.15).

In 1950, rediscovering unpublished work of von Neumann, Nachman Aronszajn showed that a compact operator on a Hilbert space always has an invariant subspace. A few years later Aronszajn and Kennan Smith generalized this result to compact operators on Banach spaces. Paul Halmos described the situation subsequent to this work as follows [18]:

> Smith pointed out, I might almost say complained, that the proof was "tight". It left no room for modifications and generalizations; it proved exactly what it was designed to prove, no more.... Aronszajn taught me the proof on a restaurant napkin several months before the paper appeared. I understood it, I cherished it, and along with many others I kept trying to "loosen" it so as to be able to apply it more broadly—but all to no avail (p. 320).

There the matter stayed until a breakthrough occured in 1966, and it was shown that any operator T on a Hilbert space $\mathscr{H}$ for which there is a nonzero polynomial $p(z) = a_n z^n + \cdots + a_1 z + a_0$ such that $p(T) = a_n T^n + \cdots + a_n T + a_0 I$ is compact (such an operator T is said to be *polynomially compact*) has an invariant subspace. The first proof of this, by Allen Bernstein and Abraham Robinson, used methods of "nonstandard analysis," but Halmos quickly reworked their argument to formulate them in classical standard analysis, publishing the resulting work as a short paper later the same year.

In 1973, the young Russian mathematician Victor Lomonosov caused a sensation by announcing a theorem which included the following result: Any operator on an infinite-dimensional complex Banach space which commutes with a nonzero compact operator has an invariant subspace. Even more, he shows that any operator which commutes with an operator (not a scalar multiple of the identity) which commutes with a nonzero compact operator has an invariant subspace. At first it was not clear whether this latter description might include all bounded linear operators. While Lomonosov's proof was short and elegant, an even briefer and more accessible proof was later provided by Hugh Hilden; the reader can find the details of Hilden's argument in [43], pp. 120–121. This particular thread of work on the invariant subspace problem—starting with von Neumann and culminating with Lomonosov and Hilden—proceeds from the philosophy that compact operators generalize finite-dimensional operators.

Another thread in the invariant subspace story is anchored by the statement that normal operators on a Hilbert space always have invariant subspaces (we'll see this in Chapter 6). One then tries to find other classes of operators, related in some way to a weakening of the normality hypothesis, which can be shown to have invariant subspaces. In particular, the class of subnormal operators (see [5]) on Hilbert spaces all have invariant subspaces.

As of this writing, the invariant subspace problem is still open for bounded linear operators on Hilbert spaces. For Banach space, though, the situation was resolved by work of Enflo published in 1987. He constructs a Banach space X and an operator in $\mathscr{B}(X)$ with no invariant subspace. While the date of publication is 1987, Enflo's announcement of the result, and a manuscript containing the example, dates from 1975. Certainly part of the explanation for the long delay before formal publication lies in the complexity of Enflo's construction. In fact, in reviewing the 100-page long *Acta Mathematica* publication, the reviewer A.M. Davie writes [9]:

[Enflo's work]...is a remarkable achievement; however the latter part of his paper is so impenetrable that it is destined to be admired rather than read.

At the same time, the basic idea that underlies Enflo's construction is a natural one in a sense that we will be better able to describe in Section 6.1. He constructs the space X as he goes, by putting a norm on the space of polynomials so that, with the resulting space completed to a Banach space, the shift operator (multiplication by the independent variable) has no invariant subspace. So here the space is complicated but the operator is simple. B. Beauzamy has published what is essentially an exposition of Enflo's example, with some considerable simplifications. Counterexamples have also been given by C. Read; one of these is an example with a simple space (ℓ^1) but a complicated operator.

Related to the question of existence of invariant subspaces is the problem of determining all of the invariant subspaces of a given operator. Generally speaking, this is a very difficult problem, although there have been some notable successes. For example, the invariant subspaces of the shift operator M_z of multiplication by z on the Hardy space H^2 have been determined and have a beautiful structure (see [40]). The operator of multiplication by z on the Bergman space $L^2_a(\mathbb{D})$ is known to have an extremely complicated lattice of invariant subspaces, and an understanding of this structure is matter of on-going work.

4.5 Introduction to the Spectrum

We start with a definition, which will be of fundamental importance to us.

Definition 4.26. If $T : X \to X$ is a bounded linear operator on a Banach space X, the set of complex numbers λ for which $T - \lambda I$ is not invertible is called the *spectrum* of T.

We will denote the spectrum of T by $\sigma(T)$. The spectrum of an operator on a Hilbert or Banach spaces contains vital information about the operator. It is an "invariant" of the operator in the sense of the following result.

Proposition 4.27. *If T is an operator in $\mathscr{B}(X)$ for a Banach space X, and if S is an invertible operator in $\mathscr{B}(X)$, then $\sigma(T) = \sigma(S^{-1}TS)$.*

Proof. If $T - \lambda I$ is invertible, with inverse V, then $S^{-1}TS - \lambda I = S^{-1}(T - \lambda I)S$ has inverse $S^{-1}VS$. Conversely, if $S^{-1}(T - \lambda I)S$ is invertible, then applying the first part we see that $S[S^{-1}(T - \lambda I)S]S^{-1} = T - \lambda I$ is invertible. Since $S^{-1}(T - \lambda I)S = S^{-1}TS - \lambda I$, this completes the proof. $\square$

When T_1 and T_2 in $\mathscr{B}(X)$ are related by $T_2 = S^{-1}T_1S$ for some invertible $S \in \mathscr{B}(X)$, we say that T_1 and T_2 are *similar*. Thus the last result says that similar operators have the same spectrum. The reader can show, by means of 2×2 matrices (that is, by operators in $\mathscr{B}(\mathbb{C}^2)$), that the converse is not true and two operators can have the same spectrum but fail to be similar.

It is helpful to think about how a complex number λ could get into $\sigma(T)$. Recall that an operator $T - \lambda I$ is invertible if and only if $T - \lambda I$ is bijective. So one way for λ to be a point of $\sigma(T)$ is for $T - \lambda I$ to fail to be one-to-one. By linearity, this happens if and only if there is a vector $g \neq 0$ with $Tg = \lambda g$, and thus λ is an eigenvalue of T.

If $\lambda \in \sigma(T)$ but λ is *not* an eigenvalue of T, then it must be the case that $T - \lambda I$ is not surjective. It sometimes helps to distinguish two ways this could happen: Either the range of $T - \lambda I$, while not all of X, is at least dense in X, or the closure of the range of $T - \lambda I$ is a proper subspace of X. These two "parts" of the spectrum are called, respectively, the continuous spectrum and the residual spectrum. In some sense this last piece is the most intractable part of the spectrum, and we'll see later that for certain classes of operators (for example, self-adjoint operators on a Hilbert space) the residual spectrum is empty. There are other useful ways of distinguishing various pieces of the spectrum; some of these (approximate eigenvalues, compression spectrum, essential spectrum) will be discussed in Sections 5.2 and 5.3.

In all of this discussion the reader should keep in mind the much simpler situation for a linear operator on a finite-dimensional space, where the operator is bijective if and only if it is injective. In other words, for an operator on a finite-dimensional space, the spectrum is just the set of eigenvalues of the operator.

Example 4.28. Consider the operator M_x of multiplication by $\varphi(x) = x$ on the Hilbert space $(L^2[0, 1], dx)$. In Exercise 4.15 you are asked to show that M_x has no eigenvalues. We claim, however, that each $0 \leq \lambda \leq 1$ is in $\sigma(M_x)$. To see this, it is helpful to recall that an invertible operator is bounded below (meaning $\|Ag\| \geq \delta \|g\|$ for some positive δ and all g; see Definition 2.24), and to observe that

$$M_x - \lambda I = M_{x-\lambda},$$

the operator of multiplication by $x - \lambda$. If $0 < \lambda < 1$ choose N sufficiently large that if $n \geq N$ then

$$E_n \equiv [\lambda - \frac{1}{n}, \lambda + \frac{1}{n}] \subseteq [0, 1]$$

and set

$$g_n = \sqrt{\frac{n}{2}} \chi_{E_n}$$

for all $n \geq N$. The g_n are unit vectors in $L^2[0, 1]$ and

$$\|(M_x - \lambda I)g_n\|^2 = \|(x - \lambda)g_n\|^2 = \frac{n}{2} \int_{E_n} |x - \lambda|^2 dx \leq \frac{n}{2} \left(\frac{1}{n}\right)^2 \frac{2}{n} = \frac{1}{n^2}.$$

This computation shows that $M_x - \lambda I$ is not bounded below, and hence not invertible. A similar argument, with $E_n = [0, 1/n]$ or $E_n = [1 - (1/n), 1]$ applies to show that $\lambda = 0$ and $\lambda = 1$ are also in $\sigma(M_x)$.

Finally, we claim that no point outside of the interval $[0, 1]$ can lie in $\sigma(M_x)$. If $\lambda \in \mathbb{C} \setminus [0, 1]$, then

$$\frac{1}{x-\lambda} \in L^\infty[0,1]$$

and $M_{1/(x-\lambda)}$ is a bounded operator which is inverse to $M_x - \lambda I$.

Many further examples of concrete operators and their spectra will appear in later sections, when we have a bit more machinery at our disposal.

The terminology "spectrum" comes from David Hilbert, who made major contributions to functional analysis initially motivated by a study of integral equations. Especially important were a collection of six papers written by Hilbert in the period 1904–1910 (and published together as a book in 1912 [19]). The fourth of this series of papers marks the beginning of the modern spectral theory. Here, in a general discussion of bilinear and quadratic forms, he generalized the concept of eigenvalue to that of the "spectrum"[2] and began the study of the relationships between the operator T and its spectrum. Retrospectively, and in modern language, we can say that he studied self-adjoint operators on ℓ^2, and an important aspect of this work was the discovery of ways to deal with the complications that arise when the continuous spectrum is not empty (and thus we are "outside" of the finite-dimensional case). This led to a description of any bounded self-adjoint operator on ℓ^2 which is a generalization of what we have done for compact self-adjoint operators in Section 4.3, and which leads to the spectral theorem as we shall discuss it in Chapter 6.

When later it was discovered that the mathematical setting of self-adjoint operators on Hilbert space was a useful mathematical tool for theoretical physicists who were developing the then new theory of quantum mechanics, the spectra of these operators became related to the explanation of the "spectra" of atoms. Hilbert comments on this remarkable coincidence of terminology [38]:

> I developed my theory of infinitely many variables from purely mathematical interests, and even called it "spectral analysis" without any presentiment that it would later find an application to the actual spectrum of physics (p. 183).

Indeed, it is remarkable how the development of the theory of operators on Hilbert spaces occurred just as it was needed for the development of quantum mechanics. As A.M. Vershik writes in an essay on functional analysis in the twentieth century [45]:

> One might even conjecture that if the functional analysis of Hilbert spaces had not yet existed at the time when quantum mechanics arose, it would have been created out of necessity. For that reason, it is no exaggeration to say that the extremely close connection between the latest physics of the first half of the twentieth century and functional analysis gave the latter even greater authority (p. 441).

---

[2] Hilbert defined the spectrum of T as the set of λ for which $I - \lambda T$ is not invertible, which gives the reciprocals of what is now the commonly used definition.

4.6 The Fredholm Alternative

Our main goal in this section is to get information on the spectrum of a *compact* operator on a Hilbert space.

Theorem 4.29. *Suppose X is a Banach space and $T \in \mathscr{B}(X)$ is compact. If $\lambda \neq 0$, then $T - \lambda I$ has closed range.*

Proof. For $\lambda \neq 0$, $T - \lambda I = \lambda(\frac{1}{\lambda}T - I)$. Since $\frac{1}{\lambda}T$ is compact if T is, it suffices to prove the theorem for $\lambda = 1$. Suppose, for a contradiction, that the range of $T - I$ is not closed. Define a map S from the quotient space $X/\ker (T - I)$ into X by $S(x + \ker (T - I)) = (T - I)x$. By Exercise 3.26 in Chapter 3, we know that S is a well-defined, bounded linear map, which is one-to-one. The range of S is equal to the range of $T - I$, and hence by our assumption, the range of S is not closed. In Exercise 2.12 of Chapter 2 it was shown that a Hilbert space operator that is bounded below must have closed range, and it is easy to see that this same result holds in Banach spaces as well. Thus S is not bounded below, and so there must be (quotient space) unit vectors $x_n + \ker (T - I)$ in $X/\ker (T - I)$ with

$$\|S(x_n + \ker (T - I))\| = \|(T - I)x_n\| \to 0.$$

By the definition of the coset norm, if $\|x_n + \ker (T - I)\| = 1$, then for any positive ε there exist $y_n \in \ker (T - I)$ such that $\|x_n - y_n\| \leq 1 + \varepsilon$. Since $\|x_n - y_n + \ker (T - I)\| = 1$, there is no loss of generality in assuming, say, that $\|x_n\| \leq 2$ for all n. Compactness of the operator T then guarantees that Tx_n has a convergent subsequence, and hence we may assume (not relabeling this subsequence) that $Tx_n \to y$ for some y in X. Since $(T - I)x_n \to 0$, we must have $x_n \to y$, and by continuity, $Tx_n \to Ty$. Thus $Ty = y$ and y is in $\ker (T - I)$. Now we have a contradiction: Writing $[x_n]$ for the coset $x_n + \ker (T - I)$, we have $\|[x_n]\| = 1$ and $\|[y]\| = 0$ but also $x_n \to y$. □

The restriction $\lambda \neq 0$ in Theorem 4.29 is crucial; see Exercise 4.26.

Theorem 4.30. *Suppose that T is a compact operator on a Hilbert space $\mathscr{H}$ and let M_j be the range of the operator $(T - I)^j$ for each $j = 1, 2, \ldots$. There exists a positive integer j such that $M_j = M_{j+1}$.*

Proof. By the previous theorem, M_1 is closed. For $j > 1$, we may expand $(I - T)^j$ by the binomial theorem to write

$$(I - T)^j = I - jT + \frac{j(j-1)}{2}T^2 + \cdots + (-1)^j T^j,$$

where $A \equiv jT - j(j-1)/2T^2 + \cdots - (-1)^j T^j$ is compact, by Proposition 4.9. Apply the previous theorem to $I - A$ to conclude that M_j is closed for each j.

Clearly $M_{j+1} \subseteq M_j$; suppose this containment is proper for each j. The quotients M_j/M_{j+1} would each then have dimension at least 1, and for each j we can choose x_j in M_j with $\|x_j + M_{j+1}\| = 1$. As in the proof of the preceding theorem, there is no loss of generality in assuming that $\|x_j\| \leq 2$ for each j. We claim that $\|Tx_j - Tx_k\| \geq 1$ for $j \neq k$, contradicting the hypothesis that T is compact. To this end, suppose $j < k$ so that $j < j+1 \leq k < k+1$. We have

- $x_k \in M_k \subseteq M_{j+1}$,
- $(T-I)x_j \in M_{j+1}$ by definition of M_{j+1}, and
- $(T-I)x_k \in M_{k+1} \subseteq M_{j+1}$.

Defining $y \equiv (T-I)x_j - (T-I)x_k - x_k$, we see that y is in M_{j+1}, and the definition of the coset norm guarantees that $\|x_j + y\| \geq 1$. But $x_j + y = Tx_j - Tx_k$, and we have verified our claim. $\square$

The result of Theorem 4.30 is sometimes phrased as "$T - I$ has finite descent if T is compact."

The next result is the main result of this section. It says that the nonzero points in the spectrum of a compact operator are always eigenvalues of the operator.

Theorem 4.31. *Suppose T is a compact operator on a Hilbert space $\mathscr{H}$ and $\lambda \neq 0$. If $T - \lambda I$ is not invertible, then λ is an eigenvalue of T.*

Proof. Since $T - \lambda I = \lambda(\frac{1}{\lambda}T - I)$, there is no loss of generality in taking $\lambda = 1$. Thus we are given that $T - I$ is not invertible. Suppose that 1 is not an eigenvalue of T. This means $\ker(T-I) = \{0\}$, and $T - I$ is one-to-one. Since it is not invertible, it must therefore fail to map *onto* $\mathscr{H}$: $(T-I)\mathscr{H}$ is properly contained in $\mathscr{H}$. Since $T - I$ is one-to-one, it follows that $(T-I)^2\mathscr{H}$ is properly contained in $(T-I)\mathscr{H}$, for if x_0 fails to be in the range of $T - I$, then $(T-I)x_0$ fails to be in the range of $(T-I)^2$. Continuing, we see that for each j, the range of $(T-I)^{j+1}$ is properly contained in the range of $(T-I)^j$. This is in contradiction to the conclusion of Theorem 4.30, and we are done. $\square$

The next result, called the "Fredholm alternative," summarizes what we have learned in this section. Notice how it captures results you know from linear algebra about linear maps on $\mathbb{C}^n$.

Theorem 4.32. *Let T be a compact operator on a Hilbert space $\mathscr{H}$. Suppose λ is a nonzero complex number.*

(a)   If $T - \lambda I$ is one-to-one, then $T - \lambda I$ is invertible.
(b)   If $T - \lambda I$ maps $\mathscr{H}$ onto $\mathscr{H}$, then $T - \lambda I$ is invertible.

Proof. The first statement is Theorem 4.31. For the second, take adjoints. If $T - \lambda I$ is onto, then $T^* - \overline{\lambda}I$ is one-to-one (by Exercise 2.16 in Chapter 2). From the first part of the theorem, $T^* - \overline{\lambda}I$ is invertible; the adjoint of its inverse provides the inverse to $T - \lambda I$. $\square$

There is a pithy way of describing the conclusions of Fredholm alternative. Think of "$(T - \lambda I)x = y$" as an equation with "y" given and "x" as the unknown. The second conclusion in the Fredholm alternative says "if a solution exists for all y, then it is unique" while the first conclusion says "if the solution is unique, it exists."

Theorem 4.32 can be extended to the following result.

Theorem 4.33. *If T is compact in $\mathcal{B}(\mathcal{H})$ and λ is a nonzero value, then*

$$\dim \ker (T - \lambda I) = \dim [\mathrm{ran}\ (T - \lambda I)]^{\perp}.$$

Theorem 4.32 is the case where the common value of the two numbers is zero. We do not give the proof of Theorem 4.33 here, but refer the interested reader to Lemma 3.2.8 in [2].

Every result in this section has an exact Banach space analogue, and only minor modifications need to be made to obtain the proofs in this more general setting. In particular Theorems 4.30 and 4.31 go through exactly as before. For Theorem 4.32, one need only pay attention to the fact that the adjoint is defined slightly differently in the Banach space context, check that it is still true that $\ker A^* = (\mathrm{ran}\ A)^{\perp}$, where now $M^{\perp}$ denotes the bounded linear functionals which are zero at each point of M, and recall that $(T - \lambda I)^* = T^* - \lambda I$.

4.7 Exercises

4.1. Suppose that $\|\cdot\|_{\alpha}$ and $\|\cdot\|_{\beta}$ are two equivalent norms on a vector space X. Show that if $(X, \|\cdot\|_{\alpha})$ is complete, then so is $(X, \|\cdot\|_{\beta})$.

4.2. Suppose that X is an n-dimensional normed linear space over $\mathbb{C}$. Show that there is a linear bijection $T : X \to \mathbb{C}^n$ such that T and T^{-1} are continuous (in your choice of a norm for $\mathbb{C}^n$); in short, every n-dimensional normed linear space over $\mathbb{C}$ is isomorphic to $\mathbb{C}^n$.

4.3. Suppose that X is a normed linear space, endowed with the metric topology, and suppose X contains a nonempty open set V such that $\overline{V}$ is compact. The goal of this problem is to show that this forces X to be finite-dimensional.

(a) Without loss of generality we may assume that $0 \in V$. Show that as x ranges over the set V, the open sets $x + \frac{1}{2}V \equiv \{x + \frac{1}{2}v : v \in V\}$ form an open cover of $\overline{V}$. By compactness, extract a finite subcover

$$\left\{ x_k + \frac{1}{2}V \right\}_{k=1}^{N}.$$

Define Y to be the span of the points $x_1, x_2, \ldots, x_N$.

(b) Show that $V \subseteq Y + \frac{1}{2^j}V$ for each positive integer j, and hence

$$V \subseteq \bigcap_{j=1}^{\infty}(Y + \frac{1}{2^j}V).$$

(c) Show that $\bigcap_1^{\infty}(Y + \frac{1}{2^j}V) = Y$.

(d) From (b) and (c) and the fact that for any $x \in X$, a sufficiently small, but nonzero multiple of x will lie in V, conclude that $X = Y$, and thus that X is finite-dimensional.

4.4. Suppose that M_1 is a closed subspace and M_2 is a finite-dimensional subspace in a normed linear space X.

(a) Show that $M_1 + M_2 \equiv \{m_1 + m_2 : m_1 \in M_1, m_2 \in M_2\}$ is a closed subspace of X. Hint: Argue that it is enough to consider M_2 to be one-dimensional, $M_2 = \{\alpha x_0 : \alpha \in \mathbb{C}\}$. Suppose $m_n + \alpha_n x_0 \to y$ where $m_n \in M_1, \alpha_n \in \mathbb{C}$. Show that $\{\alpha_n\}$ is a bounded sequence of complex numbers, and extract a convergent subsequence α_{n_k}. Write $m_{n_k} = (\alpha_{n_k} x_0 + m_{n_k}) - \alpha_{n_k} x_0$.
(b) Use (a) to give an alternate proof of the statement in Proposition 4.3 that a finite-dimensional subspace in a normed linear space is closed.

4.5. This problem provides an alternate proof to Proposition 4.4. Suppose that $T : X \to Y$ is linear, where X and Y are normed linear spaces and X is finite-dimensional. Define $\| \cdot \|_\beta$ on X by

$$\|x\|_\beta = \max(\|x\|_X, \|Tx\|_Y).$$

(a) Check that $\| \cdot \|_\beta$ is a norm on X.
(b) Argue that $T : (X, \| \cdot \|_\beta) \to Y$ is continuous, and hence that so is $T : X \to Y$ in the original norm on X.

4.6. Let X be a Banach space and suppose T_1, T_2, S are bounded linear operators from X into X, with T_1 and T_2 compact. Show that $T_1 + T_2$, αT_1, ST_1, and $T_1 S$ are all compact (α any scalar). If F is a finite rank operator, show that SF and FS are finite rank as well.

4.7. Find the error in the following "proof" that the compact operators on a Banach space X are closed in the bounded operators on X.

Alleged proof: Suppose T_n is compact for each n and suppose further that $\|T_n - T\| \to 0$ for some bounded linear operator T. To show that T is compact, we want to show that for an arbitrary bounded sequence $\{x_n\}$ in X, $\{Tx_n\}$ has a convergent subsequence. Fix such a sequence $\{x_n\}$ and let M be a bound for it: $\|x_n\| \leq M$ for all n. Now choose an $\varepsilon > 0$, and find K sufficiently large that

$$\|T_K - T\| \leq \frac{\varepsilon}{3M}.$$

We are given that the operator T_K is compact, so we can find a subsequence $\{x_{n_j}\}$ of our sequence $\{x_n\}$ so that $T_K(x_{n_j})$ converges. But a convergent sequence must be a Cauchy sequence, so if n_j and n_k are sufficiently large, say if $n_j, n_k \geq N$, then

$$\|T_K(x_{n_j}) - T_K(x_{n_k})\| < \frac{\varepsilon}{3}.$$

We claim that $T(x_{n_j})$ converges. Since we are in a Banach space, to verify this, it is enough to show that $\{T(x_{n_j})\}$ is a Cauchy sequence. To this end, notice that for $n_j, n_k \geq N$ we have

$$\|T(x_{n_j}) - T(x_{n_k})\| \leq \|T(x_{n_j}) - T_K(x_{n_j})\| + \|T_K(x_{n_j}) - T_K(x_{n_k})\|$$
$$+ \|T_K(x_{n_k}) - T(x_{n_k})\|$$
$$\leq \|T - T_K\| \cdot M + \frac{\varepsilon}{3} + \|T_K - T\| \cdot M$$
$$\leq \varepsilon.$$

This shows that $\{T(x_{n_j})\}$ is Cauchy, and hence it converges, as desired.

4.8. If $A \in \mathcal{B}(\mathcal{H})$ is a diagonal operator with diagonal $\{\alpha_n\}$, show that if A is compact, then $\lim_{n \to \infty} \alpha_n = 0$.

4.9. This problem builds on Exercise 2.6 in Chapter 2. Show that every finite rank operator T in a Hilbert space $\mathcal{H}$ can be described as

$$Th = \sum_{j=1}^{n} \langle h, x_j \rangle y_j$$

for orthonormal vectors $x_1, x_2, \ldots, x_n$ and vectors $y_1, y_2, \ldots, y_n$.

4.10. (a) Give an example of a compact operator which is not Hilbert–Schmidt. (Hint: look for a diagonal operator with this property.)

(b) Show that no compact operator on an infinite-dimensional Hilbert space is invertible. (Exercise 4.24 below extends this result to Banach spaces).

(c) Show that if T is compact in $\mathcal{B}(\mathcal{H})$, where $\mathcal{H}$ is a Hilbert space, and $\lambda \neq 0$ is an eigenvalue of T, then ker $(T - \lambda I)$ is finite-dimensional.

4.11. For $\lambda \in \mathbb{C}$ we abbreviate $T - \lambda I$ (I the identity operator) by $T - \lambda$.

(a) Suppose that T is a normal operator in $\mathcal{B}(\mathcal{H})$. Show that

$$\|(T - \lambda)h\| = \|(T - \lambda)^* h\|$$

for all h in $\mathcal{H}$ and all scalars λ. Hence ker$(T - \lambda) = \ker(T - \lambda)^*$.

(b) Show that if T is normal, then eigenvectors corresponding to distinct eigenvalues are orthogonal.

(c) State and prove a version of Theorem 4.21 for compact normal operators.

4.12. If $T \in \mathcal{B}(\mathcal{H})$ for some (complex) Hilbert space $\mathcal{H}$ and $\langle Th, h \rangle$ is real for all $h \in \mathcal{H}$, show that T is self-adjoint.

4.13. In the notation of Corollary 4.25, show that for a given $h \in \mathcal{H}$, we can solve the equation $Tf = h$ for f if and only if $h \perp \ker T$ and

$$\sum \frac{1}{\lambda_n^2} |\langle h, e_n \rangle|^2 < \infty.$$

Find all such solutions f under these assumptions.

4.14. Consider the weighted shift operator on ℓ^2 given by

$$W(x_1, x_2, x_3, \ldots) = (0, x_1, \frac{1}{2}x_2, \frac{1}{3}x_3, \ldots).$$

Show that W is compact, but W has no eigenvalues. Find a nontrivial closed invariant subspace for W.

4.15. Let M_x be the multiplication operator acting on $L^2([0,1], dx)$ by $M_x(f) = xf$. Note that M_x is self-adjoint. Show that it has no eigenvalues, but many reducing subspaces.

4.16. Show that the Volterra operator V of indefinite integration on $L^2([0,1], dx)$, defined by

$$Vf(x) = \int_0^x f(t)\,dt,$$

is compact.

4.17. Show that the subspaces

$$M_\alpha = \{f \in L^2[0,1] : f = 0 \text{ almost everywhere on } [0, \alpha]\}$$

for any $0 \le \alpha \le 1$ are invariant subspaces for the Volterra operator V (defined in Exercise 4.16).

In fact, every invariant subspace of V is of the form M_α for some α. This is a deep result; a proof can be found in [35].

4.18. Suppose that M is a closed subspace of $\mathcal{H}$ so that $\mathcal{H} = M \oplus M^\perp$. If A is in $\mathcal{B}(\mathcal{H})$, we can write A as a matrix with operator entries

$$A = \begin{bmatrix} X & Y \\ Z & W \end{bmatrix},$$

where $X \in \mathcal{B}(M)$, $Y \in \mathcal{B}(M^\perp, M)$, $Z \in \mathcal{B}(M, M^\perp)$, and $W \in \mathcal{B}(M^\perp)$. If M is an invariant subspace for A, what does this tell you about Z? If M is reducing subspace for A, what further information do you have about the operator entries of this matrix?

4.19. If $\{e_n\}$ is an orthonormal sequence in a Hilbert space $\mathcal{H}$, and $T \in \mathcal{B}(\mathcal{H})$ is compact, show that $Te_n \to 0$.

4.20. Show that there is no nonzero multiplication operator on $L^2(T, dx/(2\pi))$ that is Hilbert–Schmidt. Is there a multiplication operator on $L^2(T, dx/(2\pi))$ that is compact?

4.21. Show that if $T \in \mathcal{B}(\mathcal{H})$ is compact, and M is a closed invariant subspace of T, then the restriction of T to M is compact.

4.22. Show that a bounded linear operator T on a Hilbert space $\mathcal{H}$ which is self-adjoint and satisfies $T^2 = T$ is an orthogonal projection onto its range.

4.23. For an analytic function φ mapping the unit disk into itself, define the linear composition operator C_φ by $C_\varphi(f) = f \circ \varphi$ for f analytic on $\mathbb{D}$.

(a) Show that if C_φ is Hilbert–Schmidt on the Bergman space $L^2_a(\mathbb{D})$, then

$$\int_\mathbb{D} \frac{1}{(1 - |\varphi(z)|^2)^2} dA(z) < \infty.$$

(b) Show that if C_φ is bounded on $L^2_a(\mathbb{D})$ and

$$\int_\mathbb{D} \frac{1}{(1 - |\varphi(z)|^2)^2} dA(z) < \infty,$$

then C_φ is Hilbert–Schmidt.

(c) Give an example of a compact composition operator on $L^2_a(\mathbb{D})$.

4.24. Show that if X is an infinite-dimensional Banach space, then no bounded linear operator on X can be both compact and invertible.

4.25. Show that the result of Exercise 4.10(c) also holds for a compact operator on a Banach space.

4.26. Show that a compact operator A on a Banach space X can only have closed range if its range is finite-dimensional.

4.27. Suppose that A is a compact operator on a Banach space X. Show that if $A^2 = A$, then the range of A is finite-dimensional.

4.28. Are the Hilbert–Schmidt operators a closed subspace of $\mathscr{B}(\mathscr{H})$?

4.29. Suppose that T is a compact operator in $\mathscr{B}(\mathscr{H})$ and $\lambda \neq 0$. Show that there exists a positive integer k so that

$$\ker (T - \lambda I)^k = \ker (T - \lambda I)^{k+1}.$$

This is sometimes described as "$T - \lambda I$ has finite ascent," since for any operator $A \in \mathscr{B}(\mathscr{H})$ we have $\ker A \subseteq \ker A^2 \subseteq \ker A^3 \subseteq \cdots$.

Chapter 5
Banach and C^*-Algebras

In 1943, a paper, written by I.M. Gelfand and M. Neumark, "On the imbedding of normed rings into the ring of operators in Hilbert space," appeared (in English) in Mat. Sbornik. From the vantage point of a fifty year history, it is safe to say that the paper changed the face of modern analysis.
R. Kadison ([25], p. 21).

In this chapter, our Banach spaces will be equipped with some additional structure which comes from a multiplication operation; that is, a Banach space $\mathscr{A}$ here will permit the multiplication of two vectors. This multiplication will be required to satisfy the following properties:

(1) $a(bc) = (ab)c$
(2) $(a+b)c = ac + bc$
(3) $a(b+c) = ab + ac$
(4) $\lambda(ab) = (\lambda a)b = a(\lambda b)$

for all a, b, c in $\mathscr{A}$ and all scalars λ. Conspicuously absent from this list is any requirement of commutativity for this new multiplication operation, as well as the requirement that there be a multiplicative unit, i.e., a vector I such that $aI = Ia = a$ for all a in $\mathscr{A}$. We will impose these additional requirements (particularly the latter) from time to time, but at the moment neither is required. The terminology "complex algebra" is used for a vector space over $\mathbb{C}$ having properties (1)–(4) above; if a unit exists for the multiplication operation, we'll say the algebra is "unital."

A *Banach algebra* is a complex algebra $\mathscr{A}$ with a norm making $\mathscr{A}$ into a Banach space and satisfying

$$\|ab\| \le \|a\|\|b\|.$$

Note this norm property guarantees that multiplication, as a map from $\mathscr{A} \times \mathscr{A}$ into $\mathscr{A}$, is continuous: if $a_n \to a$ and $b_n \to b$ then $a_n b_n \to ab$. This follows by writing $a_n b_n - ab = (a_n - a)b_n + a(b_n - b)$. When $\mathscr{A}$ is unital, we assume $\|I\| = 1$; see Exercise 5.2.

The final layer of structure we will impose on some of the Banach algebras to be studied comes from the notion of an involution. An *involution*, on a Banach algebra $\mathscr{A}$, is a map $a \to a^*$ of $\mathscr{A}$ into $\mathscr{A}$ satisfying

(1) $(a^*)^* = a$
(2) $(ab)^* = b^* a^*$
(3) $(\lambda a + b)^* = \overline{\lambda} a^* + b^*$

B.D. MacCluer, *Elementary Functional Analysis*, DOI 10.1007/978-0-387-85529-5_5,   107
© Springer Science+Business Media, LLC 2009

for $a, b \in \mathscr{A}$ and λ a scalar. We call a^* the adjoint of a.

Finally, a C^*-*algebra* is a Banach algebra with an involution such that

$$\|a^* a\| = \|a\|^2.$$

We will call this the C^*-*identity*. One way to motivate the "naturalness" of this last definition is to recall that for a bounded linear operator A on a Hilbert space $\mathscr{H}$ we have already observed $\|A^* A\| = \|A\|^2$ (Proposition 2.14). It is occasionally helpful to note that a Banach algebra with involution satisfying the inequality $\|a^* a\| \geq \|a\|^2$ for all $a \in \mathscr{A}$ is a C^*-algebra, meaning we get the inequality in the other direction for free. You are asked to provide the proof for this in Exercise 5.1.

5.1 First Examples

Let us look at some examples, which show that these new definitions are all quite natural.

Example 5.1. Consider $\mathbb{C}$ with the usual multiplication, absolute value as norm, and conjugation as involution: $z^* = \bar{z}$. The C^*-identity is the familiar statement $|\bar{z}z| = |z|^2$, and $\mathbb{C}$ is a commutative C^*-algebra with unit, 1.

Example 5.2. Let X be any compact Hausdorff space, and consider the Banach space $C(X)$ of all continuous, complex-valued functions on X in the supremum norm, with pointwise-defined multiplication. This is a commutative Banach algebra, with the constant function 1 serving as the multiplicative unit. Defining an involution on $C(X)$ by $f^*(x) = \overline{f(x)}$ makes $C(X)$ into a C^*-algebra. We'll see later that every commutative unital C^*-algebra is "isometrically isomorphic" to $C(X)$ for some choice of a compact Hausdorff space X.

Example 5.3. Now let $X = \mathbb{R}$ and consider the Banach space $C_0(\mathbb{R})$ of continuous complex-valued functions that vanish at ∞ (meaning $\lim_{x \to \pm\infty} f(x) = 0$, or equivalently, that $\{x : |f(x)| \geq \varepsilon\}$ is compact for every $\varepsilon > 0$) in the supremum norm. Define multiplication pointwise and involution just as in the previous example. Then $C_0(\mathbb{R})$ is a commutative, but nonunital, C^*-algebra. This example can be generalized by replacing the real line by any noncompact but locally compact Hausdorff space X; analogously to the last comment in the previous example, every commutative nonunital C^*-algebra is $C(X)$ for some locally compact Hausdorff space X.

Example 5.4. Starting with a σ-finite measure space $(X, \mathfrak{M}, \mu)$, the Banach space $L^\infty(X, \mu)$, with multiplication and involution defined as in the last two examples, is a commutative, unital C^*-algebra. (Strictly speaking, to define the multiplication of two elements of $L^\infty(X, \mu)$ we choose a representative of each and define its pointwise product to be a representative of the product element.) Note it would not do to replace "∞" by "$p < \infty$" in this example, as the multiplication of two L^p functions need not be in L^p for finite p.

Example 5.5. Our most important example is the Banach algebra of all bounded linear operators on a Hilbert space $\mathscr{H}$, normed by the operator norm $\|A\| = \sup\{\|Ah\| : \|h\| = 1\}$, and with multiplication defined by composition $(AB)(h) = A(B(h))$. This is a noncommutative (when the dimension of $\mathscr{H}$ is at least two) Banach algebra with identity I. Defining A^* to be the usual operator adjoint provides an involution on $\mathscr{B}(\mathscr{H})$ under which we have a C^*-algebra, as noted in Chapter 2. In the special case that $\mathscr{H} = \mathbb{C}^n$, then $\mathscr{B}(\mathscr{H})$ is identified with the $n \times n$ matrices, and we will often denote this by $\mathbb{M}_n$. When $\mathscr{H}$ is replaced by a Banach space X, $\mathscr{B}(X)$ is a Banach algebra.

Even though many of the classical Banach spaces are in fact Banach algebras under a natural multiplication, the conscious exploitation of this fact was rather long in coming. Riesz, writing in 1913, looked explicitly at the product of operators on a Hilbert space, and was at least implicitly aware of the inequality $\|AB\| \leq \|A\|\|B\|$. By 1930, the concept of "rings of operators" came under explicit study, and beginning in 1936 an important series of papers by Francis Murray and John von Neumann, titled "On Rings of Operators," developed the theory of what are now called *von Neumann algebras*. These are certain kinds of C^*-subalgebras of $\mathscr{B}(\mathscr{H})$, and a particular motivation for their study was to provide the "right" mathematical framework for the study of "observables" in quantum mechanics.

Von Neumann was a brilliant and prolific mathematician who made fundamental contributions to many areas of both pure and applied mathematics. He (along with Albert Einstein and Kurt Gödel) was part of the first faculty at the Institute for Advanced Study in Princeton. Peter Lax, in the forward to a recently published collection of letters written by von Neumann, says

> ...had he lived a normal span of years[1], he would certainly have been a recipient of a Nobel Prize in economics. And if there were Nobel Prizes in computer science and mathematics, he would have been honored by these, too. So the writer of these letters should be thought of as a triple Nobel laureate, or possibly, a $3\frac{1}{2}$-fold winner, for his work in physics, in particular quantum mechanics ([37], p. xiii).

His work with Murray is among his most influential, at least on the pure mathematics side. Curiously, this work predates much of the foundational work on Banach algebras that we will look at in the next sections (much of which is due to I. Gelfand).

Von Neumann showed a prodigious talent as a young child for calculation and solving problems. According to a biographical article on von Neumann written by Halmos [16],

> At the age of 6 he could divide two eight digit numbers in his head; by 8 he had mastered the calculus; by 12 he had read and understood Borel's *Théorie des Fonctions* (p. 383).

Stories about his astonishing calculational abilities recur throughout his life. His biographer, N. Macrae, tells the following anecdote [30]:

---

[1] Von Neumann died in 1957, at the age of 53, of cancer.

When calculating a problem while sitting, he was apt to stare at the ceiling muttering, with an almost frighteningly blank face. He did this when the Rand Corporation asked whether his computers could be modified to tackle a particular problem, which—as Rand staff explained to him for two hours on blackboards and with graphs—would understandably be beyond computers in their present state. For two or three minutes, Johnny "stared so blankly that a Rand scientist later said he looked as if his mind had slipped his face out of gear. Then he said 'Gentlemen, you do not need the computer, I have the answer'" (p. 9).

Any norm-closed subalgebra of a C^*-algebra which is also closed under adjoints is again a C^*-algebra. An important example is the subalgebra of compact operators $\mathcal{K}(\mathcal{H})$ in $\mathcal{B}(\mathcal{H})$ (Theorem 4.10 and Proposition 4.12 are relevant here); it is noncommutative and has no unit when the dimension of $\mathcal{H}$ is infinite.

The term B^*-algebra also appears in the literature, for what we call a C^*-algebra. At one point in time, the terminology C^*-algebra was reserved for a closed subalgebra of $\mathcal{B}(\mathcal{H})$ which was also closed under the $*$ operation. The Gelfand–Naimark theorem established the fact that every B^*-algebra was "isometrically $*$- isomorphic" to a closed $*$-subalgebra of $\mathcal{B}(\mathcal{H})$ for some choice of a Hilbert space $\mathcal{H}$, and the need for separate terminology disappeared, with "C^*-algebra" winning out.

Since $\mathcal{B}(\mathcal{H})$ is our most important example of a C^*-algebra, and in view of the Gelfand–Naimark theorem just discussed, we will henceforth use uppercase letters $A, B, C, \ldots$ to denote elements of a generic Banach or C^*-algebra, and write I for the multiplicative unit, if there is one. We will also carry over some familiar terminology from the $\mathcal{B}(\mathcal{H})$ setting and call an element A of a C^*-algebra *self-adjoint* if $A = A^*$ and *normal* if $AA^* = A^*A$. Thus, for example, the self-adjoint elements of $C(X)$ as in Example 5.2 above are the real-valued functions in $C(X)$.

5.2 Results on Spectra

For a bit, we don't need the C^*-structure, so in this section we work in Banach algebras. Since we'll be concerned with invertibility, we will assume our Banach algebras to be unital. As noted above, the bounded linear operators $\mathcal{B}(X)$, where X is a Banach space, form a Banach algebra, under the usual multiplication of operators (that is, composition).

Definition 5.6. Suppose $\mathscr{A}$ is a Banach algebra with unit I. We say that $A \in \mathscr{A}$ is *invertible* if there exists B in $\mathscr{A}$ with $AB = BA = I$.

We could talk separately about "left" and "right" inverses for an element in a Banach algebra. It is easy (and useful!) to note that if A has both a right inverse B and left inverse C, then $B = C$, since $C = CI = C(AB) = (CA)B = IB = B$. Thus when convenient, we can show that an element is invertible by separately exhibiting a left and right inverse.

For $n \times n$ matrices, the existence of either a right or left inverse guarantees invertibility of the matrix. The analogous statement does not hold more generally, though. For example, the forward shift S on ℓ^2 has left inverse S^* but is not invertible.

In the next result, the topology in use in $\mathscr{A}$ is of course the metric topology which comes from the norm.

Theorem 5.7. *Suppose $\mathscr{A}$ is a unital Banach algebra and let $\mathscr{G}$ denote the invertible elements of $\mathscr{A}$. Then $\mathscr{G}$ is an open set in $\mathscr{A}$.*

To prove this theorem, we begin with a lemma which says that the open ball of radius 1 about the identity is contained in $\mathscr{G}$.

Lemma 5.8. *If $B \in \mathscr{A}$ and $\|I - B\| < 1$, then B is invertible, and its inverse is given by $\sum_{k=0}^{\infty}(I - B)^k$.*

Notice that the formula for B^{-1} follows formally from a geometric series type manipulation: $B^{-1} = [I - (I - B)]^{-1} = \sum_{k=0}^{\infty}(I - B)^k$, where we interpret $(I - B)^0$ to be I.

Proof (Lemma 5.8). Let $C = I - B$ so that $\|C\| = r < 1$ and $\|C^n\| \le \|C\|^n = r^n$. Since $r < 1$ we have $\sum_0^{\infty} \|C^n\| < \infty$. This says the partial sums of $\sum_0^{\infty} C^n$ form a Cauchy sequence, and hence by completeness they converge in $\mathscr{A}$. Denote $\sum_{k=0}^{n} C^k$ by Z_n and $\sum_{k=0}^{\infty} C^k$ by Z. Since $Z_n = I + C + C^2 + \cdots C^n$, we have $Z_n(I - C) = I - C^{n+1} \to I$. On the other hand, $Z_n(I - C) \to Z(I - C)$, so we must have $Z(I - C) = I$. Similarly, $(I - C)Z = I$, and $I - C$ is invertible, with inverse Z. Since $I - C = I - (I - B) = B$ we are done. $\square$

With the lemma in hand, we can make short work of the proof of Theorem 5.7.

Proof (Theorem 5.7). Suppose $A_0 \in \mathscr{G}$ with inverse B_0. We claim that for all A satisfying $\|A - A_0\| < \|B_0\|^{-1}$, A is invertible. To see this, note that $\|A - A_0\| < \|B_0\|^{-1}$ implies that $\|B_0 A - B_0 A_0\| < 1$; that is, $\|I - B_0 A\| < 1$. Invoking Lemma 5.8, we find that $B_0 A$ is invertible, say $(B_0 A)^{-1} = C$. Since $I = C(B_0 A) = (CB_0)A$, we see that A is left invertible. Similarly, $\|A - A_0\| < \|B_0\|^{-1}$ implies $1 > \|AB_0 - A_0 B_0\| = \|AB_0 - I\|$ and AB_0 is invertible, with inverse, say, D. Thus A has right inverse $B_0 D$. As we have already observed, the existence of a left inverse and a right inverse for A shows that $A \in \mathscr{G}$, and the proof is complete. $\square$

The map from $\mathscr{G}$ to $\mathscr{G}$ sending A to A^{-1} is continuous; see Exercise 5.3.

We have looked at the notion of the spectrum of an operator in Section 4.4; it is straightforward to formulate the definition more generally, for an element of a unital Banach algebra.

Definition 5.9. *Let $A \in \mathscr{A}$, where $\mathscr{A}$ is a unital Banach algebra. The* spectrum *of A, denoted $\sigma(A)$, is $\{\lambda \in \mathbb{C} : A - \lambda I \text{ is not invertible in } \mathscr{A}\}$.*

We often write $A - \lambda$ for $A - \lambda I$. Of course, when $\mathscr{A}$ is $\mathscr{B}(X)$ for some Banach space X, the above definition is precisely our earlier notion of the spectrum of a bounded linear operator. As another example, suppose that $\mathscr{A}$ is the Banach algebra is $C(X)$ for some compact Hausdorff space X. The spectrum of f in $C(X)$ is the set of complex numbers λ for which $f(x) - \lambda = 0$ for some $x \in X$, that is to say, $\sigma(f) = f(X)$, the range of f.

The next result collects some of the basic properties of the spectrum of an element in a unital Banach algebra, the deepest of which is the statement that the spectrum is always nonempty. A moment's reflection on the linear algebra roots of this statement puts this in some perspective. When A is an $n \times n$ complex matrix (i.e., an element of the Banach algebra $\mathbb{M}_n$), the spectrum of A is exactly the set of eigenvalues of A. The statement that the spectrum is nonempty becomes, in this setting, the statement that every $n \times n$ matrix has a (complex) eigenvalue, a nontrivial fact whose usual proof makes use of the fundamental theorem of algebra.

Theorem 5.10. *Suppose $\mathscr{A}$ is a unital Banach algebra and let $A \in \mathscr{A}$. The spectrum of A is a nonempty, compact subset of $\mathbb{C}$, which is contained in the closed disk $\{\lambda : |\lambda| \leq \|A\|\}$. Moreover, the map which sends z to $(z - A)^{-1}$ is an $\mathscr{A}$-valued strongly analytic function on the open set $\mathbb{C} \backslash \sigma(A)$.*

The last part of the statement of the theorem requires some explanation. How do we define analyticity for vector-valued functions? For an open set Ω in the complex plane and a Banach space $\mathscr{A}$ define the derivative of the vector-valued function $f : \Omega \to \mathscr{A}$ at $z_0 \in \Omega$ to be

$$f'(z_0) = \lim_{h \to 0} \frac{f(z_0 + h) - f(z_0)}{h}$$

if it exists in $\mathscr{A}$. The quotients $(f(z_0 + h) - f(z_0))/h$ are vectors in $\mathscr{A}$, and the limit is taken in the norm topology of $\mathscr{A}$, as h tends to 0 in $\mathbb{C}$. We say f is *strongly analytic* in Ω if f' exists and is continuous in Ω. Another natural way to contemplate defining analyticity for a Banach space-valued function is as follows: Say that $f : \Omega \to \mathscr{A}$ is *weakly analytic* if $\varphi \circ f$ is analytic in the ordinary sense for every bounded linear functional $\varphi \in \mathscr{A}^*$. It is easy to show that a Banach space-valued strongly analytic function is weakly analytic (Exercise 5.4); remarkably the converse is also true, so that these two definitions of analyticity actually coincide. Since we won't need this latter fact, we omit the proof here.

To prove Theorem 5.10 will need a "Liouville-type theorem" for vector-valued analytic functions, which we turn to next.

Theorem 5.11. *If $f : \mathbb{C} \to \mathscr{A}$ is weakly analytic, where $\mathscr{A}$ is a Banach space, and bounded in $\mathbb{C}$, then f is constant.*

Proof. We are given that $\|f(z)\| \leq M < \infty$ for all $z \in \mathbb{C}$. If φ is arbitrary in $\mathscr{A}^*$, then $\varphi \circ f$ is entire and

$$|\varphi(f(z))| \leq \|\varphi\| \|f(z)\| \leq \|\varphi\| M$$

for all $z \in \mathbb{C}$. Thus for every bounded linear functional φ on $\mathscr{A}$, $\varphi \circ f$ is a bounded entire function in the complex plane, and hence constant.

We claim this implies that f is constant. Suppose not, and find z_1, z_2 with $f(z_1) \neq f(z_2)$. By Corollary 3.4 to the Hahn–Banach theorem, $\mathscr{A}^*$ separates the points of $\mathscr{A}$ and we may find $\varphi \in \mathscr{A}^*$ with $\varphi(f(z_1)) \neq \varphi(f(z_2))$, a contradiction. This verifies the claim and completes the proof. $\qquad\square$

We're now ready to prove Theorem 5.10.

Proof (Theorem 5.10). Let $|\lambda| > \|A\|$ and write $A - \lambda = \lambda(\frac{A}{\lambda} - I)$. By Lemma 5.8 we can see that $I - A/\lambda$, and hence $A - \lambda$, is invertible. Thus the spectrum of A is contained in the closed disk $\{\lambda : |\lambda| \leq \|A\|\}$.

Next we show that $\sigma(A)$ is closed by showing its complement is open. If $A - \lambda$ is invertible, we want to show that for some $\varepsilon > 0$, $A - \mu$ is invertible provided $|\lambda - \mu| < \varepsilon$. Since the set $\mathscr{G}$ of invertible elements of $\mathscr{A}$ is open, we can find an $\varepsilon > 0$ so that if $\|B - (A - \lambda)\| < \varepsilon$, then B is invertible. For this ε, $|\lambda - \mu| < \varepsilon$ implies $A - \mu$ is invertible. This shows that the complement of $\sigma(A)$ is open. Being closed and bounded in $\mathbb{C}$, $\sigma(A)$ is compact.

Define $F : \mathbb{C} \backslash \sigma(A) \to \mathscr{A}$ by $F(\lambda) = (\lambda - A)^{-1}$. We claim that F is (strongly) analytic in $\mathbb{C} \backslash \sigma(A)$. For h in $\mathbb{C}$ sufficiently small so that $\lambda + h$ stays in the open set $\mathbb{C} \backslash \sigma(A)$ we have

$$(\lambda + h - A)^{-1} - (\lambda - A)^{-1} = (\lambda + h - A)^{-1}[(\lambda - A) - (\lambda + h - A)](\lambda - A)^{-1}$$

so that

$$\frac{F(\lambda + h) - F(\lambda)}{h} = -(\lambda + h - A)^{-1}(\lambda - A)^{-1}.$$

Continuity of the inverse (Exercise 5.3) shows that

$$\lim_{h \to 0} \frac{F(\lambda + h) - F(\lambda)}{h} = -[(\lambda - A)^{-1}]^2$$

verifying the analyticity of F.

Finally we turn to the assertion that the spectrum of A is nonempty. If it were empty, then the $\mathscr{A}$-valued function F as just defined is analytic in all of $\mathbb{C}$. Moreover, for $|\lambda| > \|A\|, F(\lambda) = (\lambda - A)^{-1} = \lambda^{-1}(I - A/\lambda)^{-1}$, which tends to 0 as $|\lambda| \to \infty$. Thus if $\sigma(A) = \emptyset$, then F is a bounded, entire $\mathscr{A}$-valued function, and hence constant by Theorem 5.11, which is clearly a contradiction. $\square$

The next result, called the Gelfand–Mazur theorem, is due independently to Gelfand (1941) and Mazur (1938). It uses Theorem 5.10 to show that the only unital Banach algebra which is also a division algebra is $\mathbb{C}$. An isomorphism between Banach algebras is a bijective linear map which is also multiplicative; it is isometric if it also preserves norms.

Theorem 5.12. *If $\mathscr{A}$ is a unital Banach algebra in which each nonzero element is invertible, then $\mathscr{A}$ is isometrically isomorphic to $\mathbb{C}$.*

Proof. Let $A \in \mathscr{A}$ and suppose λ_1, λ_2 are two distinct complex numbers. At least one of $A - \lambda_1, A - \lambda_2$ is invertible (since both can't be 0). On the other hand, $\sigma(A)$ is nonempty, so $\sigma(A)$ consists of exactly one complex number for each $A \in \mathscr{A}$; call it $\lambda(A)$. Now $A - \lambda(A)I = 0$ or $A = \lambda(A)I$. The mapping sending A to $\lambda(A)$ is an isometric isomorphism of $\mathscr{A}$ onto $\mathbb{C}$; verification of the routine details of this statement is left to the reader. $\square$

For a polynomial $p(z) = c_n z^n + \cdots + c_1 z + c_0$ and an element A of a unital Banach algebra, we write $p(A)$ for $c_n A^n + \cdots + c_1 A + c_0 I$ and $p(\sigma(A))$ for the set $\{p(\lambda) : \lambda \in \sigma(A)\}$.

Lemma 5.13. *The product $\Pi_1^n (A - \lambda_j I)$ is invertible if and only if each of the factors $A - \lambda_j I$ is invertible.*

Proof. The "if" direction of the lemma is obvious. Now suppose $\Pi_1^n (A - \lambda_j I)$ is invertible, with inverse S. Fix a j, with $1 \leq j \leq n$, and set

$$P = \Pi_{m \neq j} (A - \lambda_m I).$$

Since $A - \lambda_m I$ and $A - \lambda_k I$ commute for every m and k, we have $SP(A - \lambda_j I) = I$. This shows that $A - \lambda_j$ has a left inverse. Similarly, $(A - \lambda_j I)PS = I$ and $A - \lambda_j I$ has a right inverse. As we have observed, this guarantees that $A - \lambda_j I$ is invertible. □

A spectral mapping theorem is a statement of the form $\sigma(p(A)) = p(\sigma(A))$ for p in some class of functions and A an element of a unital Banach algebra. Our first spectral mapping theorem uses the class of polynomial functions.

Theorem 5.14. *Suppose $\mathscr{A}$ is a unital Banach algebra, A is an element of $\mathscr{A}$, and p is a polynomial. We have*

$$\sigma(p(A)) = p(\sigma(A)).$$

Proof. The result is easy when p is a constant, so we assume p has degree at least one. We'll show the two inclusions: $p(\sigma(A)) \subseteq \sigma(p(A))$ and the reverse. For the first, let $\lambda \in \sigma(A)$ and factor $p(z) - p(\lambda)$ as

$$c(z - \lambda)(z - \lambda_2) \cdots (z - \lambda_n),$$

where $c \neq 0$ and the λ_j are complex numbers; note that we have used the fact that λ is a root of $p(z) - p(\lambda)$. Defining $\lambda_1 = \lambda$, we then write

$$p(A) - p(\lambda)I = c(A - \lambda_1 I)(A - \lambda_2 I) \cdots (A - \lambda_n I).$$

The fact that $A - \lambda_j I$ and $A - \lambda_k I$ commute is being used here. By Lemma 5.13, invertibility of the product $\Pi_1^n (A - \lambda_j I)$ is equivalent to invertibility of each factor; since $\lambda_1 = \lambda \in \sigma(A)$, this says that $p(A) - p(\lambda)I$ is not invertible and hence $p(\lambda)$ is in $\sigma(p(A))$.

To see that $\sigma(p(A)) \subseteq p(\sigma(A))$, let $\mu \in \sigma(p(A))$. Factor the polynomial $p(z) - \mu$ as $c(z - \beta_1) \cdots (z - \beta_n)$ so that $p(A) - \mu I = c(A - \beta_1 I) \cdots (A - \beta_n I)$. Now Lemma 5.13 says that some $A - \beta_j I$ is not invertible, so for some j, $\beta_j \in \sigma(A)$. But $p(\beta_j) - \mu = 0$, so we have realized μ as $p(\beta_j)$ where $\beta_j \in \sigma(A)$, and the proof is complete. □

As a simple example to illustrate this spectral mapping theorem, suppose that A is an element of a unital Banach algebra that satisfies $A^2 = A$. This will imply that $\sigma(A) \subseteq \{0, 1\}$. To see that this follows directly from Theorem 5.14, let $p(z) = z^2 - z$

so that $p(A) = 0$, and hence $\sigma(p(A)) = \{0\}$. We must then have $p(\lambda) = 0$ for all $\lambda \in \sigma(A)$, and thus $\sigma(A) \subseteq \{0, 1\}$.

When A is a bounded operator on a Hilbert space $\mathcal{H}$ there are extensions of Lemma 5.13 which allow us to prove spectral mapping theorems for various "parts" of the spectrum of A. Recall that by Theorem 2.25, a bounded linear operator on a Hilbert space is invertible if and only if it is bounded below and has dense range. Thus a complex number λ is in the spectrum of $A \in \mathcal{B}(\mathcal{H})$ if and only if $A - \lambda I$ is not bounded below and/or $A - \lambda I$ does not have dense range. This tells us that $\sigma(A)$ is composed of two possibly overlapping sets, pictured below:

$$\sigma_{ap}(A) \equiv \{\lambda : A - \lambda I \text{ is not bounded below}\}$$

(called the *approximate point spectrum* of A) and

$$\Gamma(A) \equiv \{\lambda : A - \lambda I \text{ does not have dense range}\}$$

(called the *compression spectrum* of A). The approximate point spectrum consists of two disjoint pieces, the eigenvalues of A (denoted $\sigma_p(A)$) and the complement of $\sigma_p(A)$ in $\sigma_{ap}(A)$. This decomposition is schematically represented below. For further discussion of these parts of the spectrum of A, see Exercises 5.5 and 5.8.

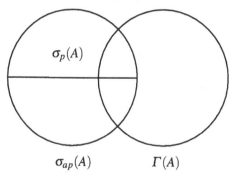

FIGURE 5.1: Parts of the spectrum

Exercise 5.7 outlines an argument to show that for every polynomial p,

$$p(\sigma_{ap}(A)) = \sigma_{ap}(p(A));$$

this is a "spectral mapping theorem" for the approximate point spectrum. Exercise 5.9 outlines a proof of an "inversion spectral mapping theorem," which says that for an invertible operator $A \in \mathcal{B}(\mathcal{H})$,

$$\sigma(A^{-1}) = [\sigma(A)]^{-1} \equiv \left\{ \frac{1}{\lambda} : \lambda \in \sigma(A) \right\}.$$

We know that the spectrum of any element A in a unital Banach algebra is compact, nonempty, and contained in the closed disk with radius $\|A\|$ centered at the origin. We define the *spectral radius* of A to be

$$r(A) \equiv \max\{|\lambda| : \lambda \in \sigma(A)\}.$$

Note that $r(A) = 0$ is equivalent to $\sigma(A) = \{0\}$, but that $r(A) = 0$ does not necessarily imply $A = 0$; it is easy to find a counterexample in $\mathbb{M}_2$.

There is a remarkable formula, due to Gelfand, for $r(A)$. This formula relates the algebraic property inherent in its definition (invertibility) to $\|A^n\|$, a metric property.

Theorem 5.15 (Spectral Radius Formula). *For $A \in \mathscr{A}$, a unital Banach algebra, $r(A) = \lim_{n \to \infty} \|A^n\|^{1/n}$.*

The *existence* of the limit is part of the proof. We will take a look at some examples and applications before proceeding to the proof.

Example 5.16. Suppose our Banach algebra is $C(X)$ for some compact Hausdorff space X. We know $\sigma(f) = \text{range } f = f(X)$, so $r(f) = \|f\|_\infty$. Thus the spectral radius formula clearly holds in this setting, since $\|f^n\|_\infty = (\|f\|_\infty)^n$ and $\|f^n\|_\infty^{1/n} = \|f\|_\infty$ for all n.

Example 5.17. If A is a self-adjoint element of a unital C^*-algebra, then we have

$$\|A^2\| = \|A^*A\| = \|A\|^2$$

and $\|A^2\| = \|A\|^2$. Since A^2 is also self-adjoint, we may replace A by A^2 to get

$$\|A^4\| = \|A^2\|^2 = \|A\|^4.$$

Continuing, an induction argument will show

$$\|A^{2^n}\| = \|A\|^{2^n},$$

and the spectral radius formula gives the conclusion that $r(A) = \|A\|$ for any self-adjoint A. Much later we will see that the same conclusion is true more generally for *normal* elements.

Suppose that $\mathscr{A}$ is a unital Banach algebra, and that $\mathscr{B}$ is closed subalgebra of $\mathscr{A}$ containing the unit I. Let B be an element of $\mathscr{B}$. It is certainly possible for the spectrum of B, relative to $\mathscr{B}$, to be different from the spectrum relative to $\mathscr{A}$. (This issue is explored further in Exercise 5.17) However, by the spectral radius formula, the spectral radius must be the same whether we think of B as an element of $\mathscr{A}$ or $\mathscr{B}$, since $\|B^n\|$ has nothing to do with which of the two algebras are being considered. In Section 5.6, we will see that if $\mathscr{A}$ is a C^*-algebra with unit I, and $\mathscr{B}$ is a C^*-subalgebra of $\mathscr{A}$ containing I, then $\sigma_{\mathscr{A}}(B) = \sigma_{\mathscr{B}}(B)$ for any $B \in \mathscr{B}$; this is referred to as "spectral permanence" in C^*-algebras.

We are ready to give the proof of the spectral radius formula.

Proof (Theorem 5.15). We begin with some observations that will clarify what we need to do. If λ is in $\sigma(A)$, then by the spectral mapping theorem (Theorem 5.14)

$\lambda^n \in \sigma(A^n)$ for any positive integer n. This tells us that $|\lambda^n| \le \|A^n\|$ and thus we have $|\lambda| \le \|A^n\|^{1/n}$ for all nonnegative integers n. It follows that

$$r(A) \le \inf_n \|A^n\|^{1/n} \le \liminf_{n \to \infty} \|A^n\|^{1/n}.$$

Thus if we can show $\limsup_{n \to \infty} \|A^n\|^{1/n} \le r(A)$ we will have

$$\limsup_{n \to \infty} \|A^n\|^{1/n} \le r(A) \le \liminf_{n \to \infty} \|A^n\|^{1/n},$$

forcing equality throughout and giving the desired result.

To this end, let Δ be the open disk in $\mathbb{C}$ centered at 0 and having radius $1/r(A)$ (which we interpret as ∞ if $r(A) = 0$). We make a few observations to get started:

- The map $z \to (z - A)^{-1}$ is (strongly) analytic in $\mathbb{C} \setminus \sigma(A)$, and has limit 0 as $|z| \to \infty$, by Theorem 5.10 and its proof.
- If λ is in $\Delta \setminus \{0\}$ then $1/\lambda$ is in $\mathbb{C} \setminus \sigma(A)$, since $|1/\lambda| > r(A)$.
- The map $\lambda \to (I - \lambda A)^{-1}$ is strongly analytic in $\Delta \setminus \{0\}$ and continuous at 0, since $(I - \lambda A)^{-1} = [\lambda(\frac{1}{\lambda} I - A)]^{-1}$.

Thus for each bounded linear functional φ in $\mathscr{A}^*$, the $\mathbb{C}$-valued function

$$f(\lambda) \equiv \varphi[(I - \lambda A)^{-1}]$$

is analytic in $\Delta \setminus \{0\}$ and continuous at 0, hence analytic in Δ. It must therefore have a power series representation in Δ, which we determine next. If

$$|\lambda| < \frac{1}{\|A\|} \le \frac{1}{r(A)},$$

we have $\|\lambda A\| < 1$ and therefore $\|I - (I - \lambda A)\| < 1$. By Lemma 5.8,

$$(I - \lambda A)^{-1} = \sum_{n=0}^{\infty} \lambda^n A^n.$$

Thus the Taylor series for $f(\lambda) = \varphi[(I - \lambda A)^{-1}]$ about 0 is

$$f(\lambda) = \sum_{n=0}^{\infty} \varphi(A^n) \lambda^n,$$

and this is valid in all of Δ. In particular, it must be the case that $\lambda^n \varphi(A^n) \to 0$ as $n \to \infty$, for each $\lambda \in \Delta$, and thus $\{\lambda^n \varphi(A^n)\}$ is a bounded sequence of complex numbers, for each fixed $\lambda \in \Delta$ and $\varphi \in \mathscr{A}^*$.

Consider the following family $\mathscr{S}$ of bounded linear functionals on $\mathscr{A}^*$:

$$\mathscr{S} = \{T_n \in \mathscr{B}(\mathscr{A}^*, \mathbb{C}) : T_n(\varphi) = \lambda^n \varphi(A^n)\},$$

where $\lambda \in \Delta$ is fixed. This is the set of elements of $\mathscr{A}^{**}$ which we previously denoted $(\lambda^n A^n)^{**}$; we know these to be bounded with norm

$$\|T_n\| = \|(\lambda^n A^n)^{**}\| = \|\lambda^n A^n\|;$$

see the discussion following Corollary 3.5. Now for each fixed $\varphi \in \mathscr{A}^*$, the fact that $\{\lambda^n \varphi(A^n)\}_{n=1}^\infty$ is a bounded sequence tells us that

$$\sup_n \{|T_n(\varphi)|\} = \sup_n \{|\lambda^n \varphi(A^n)|\} < \infty,$$

or in other words, the sequence of bounded linear functionals T_n is pointwise bounded. By the uniform boundedness principle, $\sup_n \|T_n\| < \infty$ and we have

$$\sup_n \|\lambda^n A^n\| \equiv M(\lambda) < \infty,$$

where $M(\lambda)$ denotes a finite constant, which may depend on $\lambda \in \Delta$. Thus for every n,

$$\|A^n\|^{1/n} \leq \frac{M(\lambda)^{1/n}}{|\lambda|}.$$

Since $M(\lambda)^{1/n} \to 1$ as $n \to \infty$,

$$\limsup_{n \to \infty} \|A^n\|^{1/n} \leq \frac{1}{|\lambda|}.$$

This holds for each $\lambda \in \Delta$, which has radius $1/r(A)$, and we conclude

$$\limsup_{n \to \infty} \|A^n\|^{1/n} \leq r(A).$$

As we previously observed, this is sufficient to complete the proof. $\qquad\qquad\square$

As an easy application of the spectral radius formula, note that if A and B are *commuting* elements in a unital Banach algebra, then Theorem 5.15 quickly shows that $r(AB) \leq r(A)r(B)$; this property is called submultiplicativity. It is also true that the spectral radius is subadditive, meaning $r(A+B) \leq r(A) + r(B)$ for *commuting* elements A and B. While this can be proved from Theorem 5.15, we will see an easier proof later (see Exercise 5.34).

For an operator T on a Hilbert space $\mathscr{H}$, we have discussed the decomposition of $\sigma(T)$ into the (possibly overlapping) subsets $\sigma_{ap}(T)$ and $\Gamma(T)$, the approximate point spectrum and compression spectrum, respectively. Recall from Section 4.5 we had a different perspective on "parts" of $\sigma(T)$. There we considered that the simplest way for λ to be in $\sigma(T)$ is if $T - \lambda I$ is not one-to-one; i.e., λ is an eigenvalue of T. The collection of the eigenvalues of T is called the point spectrum of T (for reasons that will become clearer in Chapter 6); recall we denote it $\sigma_p(T)$. If λ is in $\sigma(T)$ but is not an eigenvalue, then the range of $T - \lambda I$ is a proper subset of $\mathscr{H}$, and this can happen in two different ways: Either the range of $T - \lambda I$ is a proper, but dense, subset of $\mathscr{H}$, or the closure of the range of $T - \lambda I$ is a proper closed subspace of $\mathscr{H}$. This leads to a classification of $\sigma(T)\backslash\sigma_p(T)$ into two disjoint pieces: the *continuous spectrum*, where the range of $T - \lambda I$ is dense in, but not equal to, $\mathscr{H}$, and the *residual spectrum*, where the closure of the range of $T - \lambda I$ is a proper subspace of $\mathscr{H}$. So now we have decomposed $\sigma(T)$ into three disjoint pieces:

$$\text{point spec.} \equiv \{\lambda : T - \lambda \text{ is not one-to-one}\}$$
$$\text{continuous spec.} \equiv \{\lambda : T - \lambda \text{ is one-to-one}, (T - \lambda)\mathscr{H} \neq \mathscr{H}, \overline{(T - \lambda)\mathscr{H}} = \mathscr{H}\}$$
$$\text{residual spec.} \equiv \{\lambda : T - \lambda \text{ is one-to-one and } \overline{(T - \lambda)\mathscr{H}} \neq \mathscr{H}\}.$$

In Figure 5.2 we redraw Figure 5.1, showing these three disjoint pieces. The point spectrum, $\sigma_p(T)$, is shown dotted. The continuous spectrum is $\sigma_{ap}(T)\backslash(\Gamma(T) \cup \sigma_p(T))$ (shown dashed), and the residual spectrum is $\Gamma(T)\backslash\sigma_p(T)$. Certain classes of operators cannot have any residual spectrum; for example self-adjoint operators (see Exercise 5.25), or more generally, normal operators.

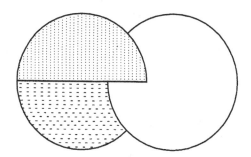

FIGURE 5.2: Disjoint parts of the spectrum

Example 5.18. Consider a (bounded) diagonal operator T on ℓ^2 with diagonal $(\alpha_1, \alpha_2, \ldots)$. Clearly each scalar α_j is an eigenvalue of T. We leave it to the reader to check that the closure of the set $\{\alpha_j\}$ is $\sigma(T)$. Suppose that $\lambda \in \sigma(T)\backslash\sigma_p(T)$. The range of the operator $T - \lambda I$ is the range of the diagonal operator S with diagonal $(\alpha_1 - \lambda, \alpha_2 - \lambda, \ldots)$. We claim that the range of S is dense in ℓ^2. To see this, note that given any point $x = (x_1, x_2, \ldots)$ in ℓ^2 and any positive integer N, the sequence

$$\left\{\frac{x_1}{\alpha_1 - \lambda}, \frac{x_2}{\alpha_2 - \lambda}, \cdots, \frac{x_N}{\alpha_N - \lambda}, 0, 0, \cdots\right\}$$

is in ℓ^2 and is mapped by S to

$$(x_1, x_2, \ldots, x_N, 0, 0, \ldots).$$

by choosing N sufficiently large, this is as close to x as desired. We have shown that every point of $\sigma(T)\backslash\sigma_p(T)$ is in the continuous spectrum, and the residual spectrum is empty.

The terms "point spectrum," "continuous spectrum," and "residual spectrum," have the same definitions for operators on a Banach space. The diagonal operator considered above, acting on the Banach space ℓ^∞, has all points $\sigma(T) \setminus \sigma_p(T)$ in the residual spectrum. The reader is asked to verify this in Exercise 5.10.

5.3 Ideals and Homomorphisms

In this section we continue to be primarily interested in the unital Banach algebra setting.

Definition 5.19. A *complex homomorphism* on a Banach algebra $\mathscr{A}$ is a linear map $\varphi : \mathscr{A} \to \mathbb{C}$ that preserves multiplication: $\varphi(AB) = \varphi(A)\varphi(B)$.

As an easy example, note that point evaluation at any $x \in X$ is a complex homomorphism on $\mathscr{A} = C(X)$ for X compact and Hausdorff.

We don't a priori require that a complex homomorphism be continuous, but we will see shortly that it automatically is, and moreover if we exclude the case that φ is identically 0, then we must have $\|\varphi\| = 1$. This nontriviality assumption also forces $\varphi(I) = 1$, since $\varphi(I) = \varphi(I \cdot I) = \varphi(I)\varphi(I)$ so that either $\varphi(I) = 1$ or $\varphi(I) = 0$; in the latter case $\varphi \equiv 0$. If A is invertible in $\mathscr{A}$ and φ is a nontrivial complex homomorphism, then $\varphi(A^{-1}) = 1/\varphi(A)$; in particular $\varphi(A) \neq 0$ if A is invertible.

Theorem 5.20. *Every nontrivial complex homomorphism φ of a unital Banach algebra is continuous and satisfies $\|\varphi\| = 1$.*

Proof. If $A \in \mathscr{A}$ with $\varphi(A) \neq 0$, then

$$\varphi\left(I - \frac{A}{\varphi(A)}\right) = 0.$$

As we have just observed, this shows that $I - A/\varphi(A)$ is not invertible, and so by Lemma 5.8 we must have $\|A/\varphi(A)\| \geq 1$, or equivalently $\|A\| \geq |\varphi(A)|$. Since $\|A\| \geq |\varphi(A)|$ clearly also holds when $\varphi(A) = 0$, we see that φ is bounded with $\|\varphi\| \leq 1$. Consideration of the identity I gives $\|\varphi\| = 1$. $\square$

A complex homomorphism is a linear and multiplicative map of a Banach algebra into the Banach algebra $\mathbb{C}$. More generally we could consider a homomorphism of a Banach algebra $\mathscr{A}$ into another Banach algebra $\mathscr{B}$; this is just a linear map that also preserves multiplication. Our primary interest will be in the case that $\mathscr{A}$ and $\mathscr{B}$ are C^*-algebras and in this case we will want the homomorphism to be star-preserving as well. Thus we define a *-homomorphism* between C^*-algebras to be a linear, multiplicative map ρ for which $\rho(A^*) = \rho(A)^*$ for all A in the domain C^*-algebra. If ρ is bijective from $\mathscr{A}$ onto $\mathscr{B}$, then ρ is a *-isomorphism* and it will preserve the C^*-structure; that is, $\mathscr{A}$ and $\mathscr{B}$ are "the same" as C^*-algebras when they are *-isomorphic. The next result says that every complex homomorphism of a C^*-algebra is automatically a *-homomorphism.

Theorem 5.21. *Suppose $\mathscr{A}$ is a unital C^*-algebra and $\varphi : \mathscr{A} \to \mathbb{C}$ is a homomorphism. For every $A \in \mathscr{A}$, $\varphi(A^*) = \overline{\varphi(A)}$.*

A consequence of the theorem is that $\varphi(A)$ is real when A is self-adjoint. This observation motivates the proof, which focuses first on self-adjoint elements.

Proof. First suppose that A is a self-adjoint element of the C^*-algebra $\mathscr{A}$ and let t be a real number. Set $B = A + itI$ so that $B^* = A - itI$ and $B^*B = A^2 + t^2I$. Since φ has norm 1, we have

$$|\varphi(B)|^2 \leq \|B\|^2 = \|B^*B\| = \|A^2 + t^2I\| \leq \|A\|^2 + t^2,$$

where we have used the C^*-identity. Setting $\varphi(A) = x + iy$ for x and y real, and recalling that $\varphi(I) = 1$, the above computation says $\|A\|^2 \geq x^2 + y^2 + 2yt$ for every real number t. Since the left-hand side is fixed, but t is arbitrary in $\mathbb{R}$, this forces $y = 0$, and we conclude that $\varphi(A)$ is real when A is self-adjoint.

The general result now follows from a standard procedure. Write an arbitrary $A \in \mathscr{A}$ as $A = X + iY$ where X and Y are self-adjoint elements of $\mathscr{A}$ (see Exercise 5.2). Thus $\overline{\varphi(A)} = \varphi(X) - i\varphi(Y)$, since we know that $\varphi(X)$ and $\varphi(Y)$ are real by the first part of the argument. Since $\varphi(A^*) = \varphi(X^*) - i\varphi(Y^*) = \varphi(X) - i\varphi(Y)$, we are done. $\square$

Homomorphisms of Banach algebras are closely connected to the notion of ideals in an algebra; we define this next.

Definition 5.22. A (two-sided) *ideal* $\mathscr{J}$ in a Banach algebra $\mathscr{A}$ is a subspace of $\mathscr{A}$ with the property that $A \in \mathscr{A}$ and $S \in \mathscr{J}$ implies $AS \in \mathscr{J}$ and $SA \in \mathscr{J}$.

At the moment we don't require that the subspace be (topologically) closed in $\mathscr{A}$. Of course, one could also talk about "left ideal" or "right ideals" in $\mathscr{A}$, by only requiring that multiplication on the left or right by arbitrary elements of $\mathscr{A}$ keeps you in $\mathscr{J}$. Unless we say explicitly otherwise, "ideal" will always mean two-sided ideal. If $\mathscr{J}$ is an ideal which is not all of $\mathscr{A}$, we will call it a *proper ideal*, and if there is no proper ideal $\mathscr{J}'$ with $\mathscr{J} \subseteq \mathscr{J}'$ but $\mathscr{J} \neq \mathscr{J}'$ we say $\mathscr{J}$ is a *maximal ideal*. An easy, but important, observation is that no proper ideal in a unital Banach algebra can contain an invertible element (else it contains I, and hence everything).

For example, in the Banach algebra $C[0,1]$ in the supremum norm, for each closed set $E \subseteq [0,1]$,

$$\mathscr{J}_E \equiv \{f \in C[0,1] : f(x) = 0 \text{ for } x \in E\} \quad .$$

is an ideal (which is moreover a closed ideal). In the next section we will see that when E is a singleton in $[0,1]$, the corresponding ideal is maximal.

The next few results collect some simple facts about ideals.

Proposition 5.23. *In a unital Banach algebra, the closure of a proper left, right, or two-sided ideal is a proper left, right, or two-sided ideal, respectively.*

Proof. We give the proof for the case of a two-sided ideal and leave the other cases for the reader. Suppose $\mathscr{J}$ is a two-sided proper ideal in $\mathscr{A}$ and let $\mathscr{G}$ be the set of invertible elements of $\mathscr{A}$. We know (Theorem 5.7) that $\mathscr{G}$ is open and nonempty; furthermore $\mathscr{J} \subset \mathscr{A} \backslash \mathscr{G}$, a closed set. In particular, $\overline{\mathscr{J}} \subset \mathscr{A} \backslash \mathscr{G}$. It is easy to check that $\overline{\mathscr{J}}$ is an ideal, and thus $\overline{\mathscr{J}}$ is an ideal not equal to $\mathscr{A}$, i.e., a proper ideal. $\square$

Proposition 5.23 is not true in the nonunital setting. For example, the non-unital Banach algebra $C_0(\mathbb{R})$ contains a proper dense ideal $C_c(\mathbb{R})$ of continuous functions with compact support; this is the statement from real analysis that the continuous functions with compact support on the real line are uniformly dense in the continuous functions which vanish at infinity.

Proposition 5.24. *In a unital Banach algebra, every maximal ideal is closed and every proper ideal is contained in a maximal ideal.*

Proof. The first statement follows from the previous result: if $\mathscr{J}$ is a nonclosed ideal, then $\overline{\mathscr{J}}$ is a proper ideal strictly containing $\mathscr{J}$, and hence $\mathscr{J}$ cannot be maximal. For the second statement, we look at the collection $\mathscr{P}$ of all proper ideals containing a given proper ideal $\mathscr{J}$, and partial order $\mathscr{P}$ by inclusion. A totally ordered chain in $\mathscr{P}$ has an upper bound (the union of all elements in the chain, which contains $\mathscr{J}$ but does not contain the identity I), so by Zorn's lemma there is a maximal element in $\mathscr{P}$. This is a maximal ideal containing $\mathscr{J}$. $\square$

We now look at the basic construct of forming a quotient by a proper, closed ideal $\mathscr{J}$ in a Banach algebra $\mathscr{A}$. Define, as for Banach spaces, the quotient space $\mathscr{A} / \mathscr{J} = \{A + \mathscr{J} : A \in \mathscr{A}\}$. Since $\mathscr{J}$ is closed, we know from Exercise 3.25 in Chapter 3 that $\mathscr{A} / \mathscr{J}$ is a Banach space under the norm

$$\|A + \mathscr{J}\| = \inf\{\|A + B\| : B \in \mathscr{J}\} = \inf\{\|A - B\| : B \in \mathscr{J}\}.$$

To put a multiplicative structure on $\mathscr{A} / \mathscr{J}$, define $(A + \mathscr{J})(B + \mathscr{J}) = AB + \mathscr{J}$. One needs to check that this multiplication is well-defined (see Exercise 5.13); here the fact that $\mathscr{J}$ is an ideal, and not merely a subspace, is important. We'd like to know that with these definitions, $\mathscr{A} / \mathscr{J}$ becomes a Banach algebra; this is the content of the next result.

Theorem 5.25. *For a Banach algebra $\mathscr{A}$ with proper closed ideal $\mathscr{J}$, the quotient $\mathscr{A} / \mathscr{J}$ is a Banach algebra. When $\mathscr{A}$ is unital, so is $\mathscr{A} / \mathscr{J}$. The quotient map $\Pi : \mathscr{A} \to \mathscr{A} / \mathscr{J}$ is a surjective homomorphism with kernel $\mathscr{J}$.*

Proof. We already know that $\mathscr{A} / \mathscr{J}$ is a Banach space. It is easy to check that the multiplication has the desired associative and distributive properties. We need to verify that

$$\|(A + \mathscr{J})(B + \mathscr{J})\| \le \|A + \mathscr{J}\|\|B + \mathscr{J}\|. \tag{5.1}$$

For any $J_1, J_2 \in \mathscr{J}$ we have

$$(A + J_1)(B + J_2) = AB + J_1 B + A J_2 + J_1 J_2,$$

where $J_1B + AJ_2 + J_1J_2$ is in $\mathscr{J}$. Thus the left hand side of Equation (5.1), which is $\|AB + \mathscr{J}\|$, is less than or equal to

$$\|AB + J_1B + AJ_2 + J_1J_2\| = \|(A + J_1)(B + J_2)\| \leq \|A + J_1\|\|B + J_2\|,$$

for any choice of J_1, J_2 in $\mathscr{J}$. If we take the infimum over all J_1, J_2 in $\mathscr{J}$, we see that (5.1) holds.

If $\mathscr{A}$ has unit I, then $\mathscr{A}/\mathscr{J}$ has unit $I + \mathscr{J}$. Note that $\|I + \mathscr{J}\| = 1$. Indeed, since 0 is in $\mathscr{J}$, $\|I + \mathscr{J}\| \leq 1$, and there cannot exist $K \in \mathscr{J}$ with $\|I - K\| < 1$, otherwise K would be invertible and $\mathscr{J}$ would not be proper.

As regards the quotient map $\Pi : \mathscr{A} \to \mathscr{A}/\mathscr{J}$, we know that Π is a bounded linear map with $\|\Pi(A)\| \leq \|A\|$ for all A. It is clear that it is surjective and its kernel is $\mathscr{J}$; the fact that it is multiplicative follows immediately from the definition of multiplication in $\mathscr{A}/\mathscr{J}$. □

Can we add C^*-structure to our quotient? Specifically, if $\mathscr{A}$ is a C^*-algebra, can we define an involution on the quotient $\mathscr{A}/\mathscr{J}$ to make it a C^*-algebra? We won't tackle this question here, except to make some observations about what the issues are. If we are able to show that closed ideals in a C^*-algebra are self-adjoint (meaning $\mathscr{J}^* = \mathscr{J}$) then setting $(A + \mathscr{J})^* = A^* + \mathscr{J}$ will be well-defined, and is readily verified to be an involution on the Banach algebra $\mathscr{A}/\mathscr{J}$. One must then check that the C^*-identity holds:

$$\|A + \mathscr{J}\|^2 = \|(A^* + \mathscr{J})(A + \mathscr{J})\|$$

for all A in $\mathscr{A}$. This takes some work, and since we will not need it in any of what follows, we refer the interested reader to [11] for a proof.

Note that the ideal $\mathscr{K} = \mathscr{K}(\mathscr{H})$ of compact operators in the bounded operators on a Hilbert space has this "self-adjoint" property just described: if $T \in \mathscr{B}(\mathscr{H})$ is compact, so is T^*, by Proposition 4.12. Suppose we form the quotient $\mathscr{B}(\mathscr{H})/\mathscr{K}(\mathscr{H})$, called the *Calkin algebra*, and consider this simply as a Banach algebra, with unit $I + \mathscr{K}$. We can ask about the spectrum of an element $T + \mathscr{K}$ in this algebra:

$$\lambda \in \sigma(T + \mathscr{K}) \iff (T - \lambda I) + \mathscr{K} \text{ is not invertible in } \mathscr{B}(\mathscr{H})/\mathscr{K}(\mathscr{H}).$$

Thus $\lambda \in \sigma(T + \mathscr{K})$ if and only if there is no $S \in \mathscr{B}(\mathscr{H})$ with $I - S(T - \lambda I)$ and $I - (T - \lambda I)S$ both compact. We call the spectrum of $T + \mathscr{K}$ the *essential spectrum* of T and denote it $\sigma_e(T)$. One sense of this terminology is that the essential spectrum of an operator is unchanged under "compact perturbations" of the operator: $\sigma_e(T) = \sigma_e(T + K)$ whenever K is a compact operator. More generally, the "essential" properties of an operator $T \in \mathscr{B}(\mathscr{H})$ are those of its image in the Calkin algebra. Thus the essential norm of T, $\|T\|_e$, is the norm of $T + \mathscr{K}$ in the Calkin algebra; that is, $\|T\|_e$ is the distance from the operator T to the ideal $\mathscr{K}(\mathscr{H})$. The essential spectral radius of T is the maximum modulus of points in the essential spectrum of T. By Theorem 5.15, the essential spectral radius is given by

$$r_e(T) = \lim_{n \to \infty} (\|T^n\|_e)^{\frac{1}{n}}.$$

An operator T is termed "essentially normal" if the coset $T + \mathscr{K}$ is normal in the Calkin algebra. Notice that this is equivalent to the requirement that $T^*T - TT^*$ be compact. Essential invertibility of T is the requirement that T be "invertible modulo the compacts." This concept will be explored more fully in Section 5.8.

Returning to the general Banach algebra setting, it is easy to see that the kernel of a complex homomorphism $\varphi : \mathscr{A} \to \mathbb{C}$ is a closed ideal in $\mathscr{A}$. The fact that $\ker \varphi$ is an ideal comes from the linear and multiplicative properties of φ. It is closed since it is the inverse image of 0 under the continuous function φ.

To get at a deeper understanding of ideals in Banach algebras, and in particular maximal ideals, we will restrict our attention for a bit to commutative Banach algebras, where we will find a rich theory. This is the subject of the next section.

5.4 Commutative Banach Algebras

Recall that a complex homomorphism of a unital commutative Banach algebra is a nontrivial (not identically 0) *multiplicative linear functional* $\varphi : \mathscr{A} \to \mathbb{C}$, which is necessarily continuous and has norm 1. We will denote the collection of all such nontrivial multiplicative linear functions by $\mathscr{M}_{\mathscr{A}}$, and call this set (for reasons that will soon become clear) the *maximal ideal space* of $\mathscr{A}$. At the moment its just a set, with no structure, although we will eventually put a useful topology on it. It is not a linear space.

The next theorem is key, and explains our terminology.

Theorem 5.26. *In a unital commutative Banach algebra $\mathscr{A}$, for every φ in $\mathscr{M}_{\mathscr{A}}$, the kernel of φ is a maximal ideal of $\mathscr{A}$, and conversely, every maximal ideal in $\mathscr{A}$ is the kernel of some $\varphi \in \mathscr{M}_{\mathscr{A}}$.*

Proof. First suppose $\mathscr{M}$ is a maximal ideal in $\mathscr{A}$, so that $\mathscr{M}$ is closed and $\mathscr{A}/\mathscr{M}$ is a unital Banach algebra. We show that this quotient is isomorphic to $\mathbb{C}$ by showing that every nonzero element is invertible and invoking Theorem 5.12, the Gelfand–Mazur theorem.

Pick $X \in \mathscr{A}$, with X not in $\mathscr{M}$; we wish to show that $X + \mathscr{M}$ is invertible in the quotient algebra. Set $\mathscr{J} = \{AX + Y : A \in \mathscr{A}, Y \in \mathscr{M}\}$. It is easy to see that this is an ideal that properly contains $\mathscr{M}$. Since $\mathscr{M}$ is maximal, we must have $\mathscr{J} = \mathscr{A}$, and thus there are elements $A \in \mathscr{A}$ and $Y \in \mathscr{M}$ with $AX + Y = I$. This says $X + \mathscr{M}$ is invertible, with inverse $A + \mathscr{M}$. By the Gelfand–Mazur theorem we know that $\mathscr{A}/\mathscr{M}$ is isometrically isomorphic to $\mathbb{C}$, via an isomorphism which we denote by i. The quotient map $\Pi : \mathscr{A} \to \mathscr{A}/\mathscr{M}$ is a homomorphism, so that the composition $i \circ \Pi$ is a complex homomorphism of $\mathscr{A}$. Its kernel is easily seen to be $\mathscr{M}$.

For the converse direction, let φ be a multiplicative linear functional on $\mathscr{A}$. We have already observed that its kernel is a closed ideal in $\mathscr{A}$. The maximality of this ideal follows from Exercise 3.26 in Chapter 3, since it says that the dimension of $\mathscr{A}/\ker \varphi$ is one.   □

Thus the map which sends $\varphi \in \mathscr{M}_{\mathscr{A}}$ to its kernel maps onto the set of maximal ideals in $\mathscr{M}$. It is easy to see that this mapping is also one-to-one. If φ_1 and φ_2 are two multiplicative linear functionals on $\mathscr{A}$ with the same kernel, we claim $\varphi_1 = \varphi_2$. We have $\varphi_1(I) = 1 = \varphi_2(I)$. Let A be arbitrary in $\mathscr{A}$. Setting $\varphi_1(A) = \alpha$ we have $\varphi_1(A - \alpha I) = 0$, and thus $A - \alpha I$ must be in the kernel of φ_2. This says $\varphi_2(A) = \alpha$. Since A was arbitrary, we conclude that $\varphi_2 = \varphi_1$. The correspondence between nontrivial multiplicative linear functionals and maximal ideals is thus one-to-one.

We now look at an example, which will suggest our next major theorem.

Example 5.27. Suppose $\mathscr{A} = C(X)$ for some compact Hausdorff space X. We claim that every nontrivial multiplicative linear functional on $C(X)$ is an evaluation functional, i.e. has the form $ev_x(f) = f(x)$ for some $x \in X$. This is equivalent to the statement that the maximal ideals in $C(X)$ all have the form

$$\mathscr{M}_x \equiv \{f \in C(X) : f(x) = 0\}$$

for some fixed $x \in X$. Since we already know that the evaluation functionals are multiplicative linear functionals with kernel $\mathscr{M}_x$, it suffices to show that every proper ideal is contained in at least one $\mathscr{M}_x$. Assume, for a contradiction, that we have a proper ideal $\mathscr{J}$ so that for each $x \in X$, we may find $f \in \mathscr{J}$ with $f(x) \neq 0$. By continuity, f is then nonzero on an open set containing x. Thus, associating to each x in X such a function f and open set $\mathscr{O}$ we get an open cover of the compact set X, and we may find a finite number of points $x_1, x_2, \ldots, x_n$, with corresponding functions f_j and open sets $\mathscr{O}_j$ covering X such that $f_j(y) \neq 0$ for all $y \in \mathscr{O}_j$. Now $\sum_1^n f_j \bar{f_j}$ is in $\mathscr{J}$ and is positive on X, hence invertible in $C(X)$. Since a proper ideal cannot contain an invertible element, we have our desired contradiction.

Now we know that the spectrum of an element f in the Banach algebra $C(X)$ is precisely the range of f, $f(X) = \{f(x) : x \in X\}$. Since the multiplicative linear functionals are the evaluation functionals, this says $\sigma(f) = \{\varphi(f) : \varphi \in \mathscr{M}_{\mathscr{A}}\}$. The next remarkable theorem says that exactly the same description holds for an arbitrary element of any commutative unital Banach algebra.

Theorem 5.28. *Suppose $\mathscr{A}$ is a commutative unital Banach algebra and let $A \in \mathscr{A}$. We have*

$$\sigma(A) = \{\varphi(A) : \varphi \in \mathscr{M}_{\mathscr{A}}\}.$$

Proof. Fix $A \in \mathscr{A}$ and suppose $\lambda \in \sigma(A)$. Then $A - \lambda I$ is not invertible and

$$\{(A - \lambda I)B : B \in \mathscr{A}\}$$

is a proper ideal (it can't contain I), and hence is contained in a maximal ideal, which by Theorem 5.26 is the kernel of some multiplicative linear functional φ. Since $\varphi(A - \lambda I) = 0$, we have $\lambda = \varphi(A)$.

Conversely, if λ is not in $\sigma(A)$, find B so that $(A - \lambda I)B = I$. Given any nontrivial multiplicative linear functional φ on $\mathscr{A}$, $\varphi(A - \lambda I)\varphi(B) = \varphi(I) = 1$ so that we must have $\varphi(A - \lambda I) \neq 0$, or $\varphi(A) \neq \lambda$ for all $\varphi \in \mathscr{M}_{\mathscr{A}}$.   □

We can read off some quick corollaries to the last theorem.

Corollary 5.29. *An element A in a unital commutative Banach algebra $\mathscr{A}$ is invertible if and only if $\varphi(A) \neq 0$ for all $\varphi \in \mathscr{M}_{\mathscr{A}}$. Furthermore, A is invertible if and only if A lies in no proper ideal of $\mathscr{A}$.*

Proof. The first statement follows immediately from the theorem. For the second statement, notice that if A is not invertible, $\{AB : B \in \mathscr{A}\}$ is a proper ideal containing A, and, as previously observed, no proper ideal contains an invertible element.   □

Corollary 5.30. *For every A in a unital commutative Banach algebra $\mathscr{A}$, and every $\varphi \in \mathscr{M}_{\mathscr{A}}$, $|\varphi(A)| \leq r(A) \leq \|A\|$.*

Proof. Only the first inequality is new, and it is an immediate consequence of Theorem 5.28.   □

Corollary 5.29 says that every nonzero noninvertible element A in a unital commutative Banach algebra sits in a proper ideal (namely $\{BA : B \in \mathscr{A}\}$) and hence in a maximal ideal, which is the kernel of a nontrivial multiplicative linear functional. In the case that the only noninvertible element is 0, we know by Theorem 5.12 that $\mathscr{A}$ is isometrically isomorphic to $\mathbb{C}$, and this isomorphism is a nontrivial complex homomorphism on $\mathscr{A}$. Commutativity is essential here; there are unital Banach algebras with no nontrivial ideals, see Exercise 5.22.

We can combine the results of Theorems 5.28 and 5.21 to obtain information on the spectrum of any self-adjoint element of a commutative unital C^*-algebra.

Theorem 5.31. *If $\mathscr{A}$ is a commutative unital C^*-algebra and $A \in \mathscr{A}$ is self-adjoint, then $\sigma(A)$ is contained in the real line.*

Proof. If $\lambda \in \sigma(A)$, then by Theorem 5.28 $\lambda = \varphi(A)$ for some $\varphi \in \mathscr{M}_{\mathscr{A}}$. Using Theorem 5.21 we have

$$\overline{\lambda} = \overline{\varphi(A)} = \varphi(A^*) = \varphi(A) = \lambda$$

since A is self-adjoint.   □

The same conclusion holds in unital C^*-algebras that are not commutative; see Theorem 5.49 below.

Right now $\mathscr{M}_{\mathscr{A}}$ is just a set. We want to put a topology on it. Since we know each $\varphi \in \mathscr{M}_{\mathscr{A}}$ is a (multiplicative) linear functional of norm 1, we can think of $\mathscr{M}_{\mathscr{A}}$ as sitting inside the norm-closed unit ball of $\mathscr{A}^*$, the dual space. Thus we might expect to use the norm-topology, i.e., the topology it inherits as a subset of $\mathscr{A}^*$, but this turns out to not be the most useful choice. In the next section we discuss the "weak*" topology (on the dual of a Banach space) and weak topologies in general; with this new notion in hand we will then return to our study of $\mathscr{M}_{\mathscr{A}}$.

5.5 Weak Topologies

The basic idea for weak topologies is as follows. Start with a Banach space X and a vector space Y of linear functionals on X rich enough to separate the points of X; that is, given $x_1 \neq x_2$ in X, there exists φ in Y with $\varphi(x_1) \neq \varphi(x_2)$. The Y-weak topology on X is defined to be the weakest (or coarsest) topology (having the smallest collection of open sets) for which all functions in Y are continuous. That is, we want $\varphi : X \to \mathbb{C}$ to be continuous for each φ in Y, so that, for example, each of the sets $\{\varphi^{-1}(U) : \varphi \in Y \text{ and } U \text{ is open in } \mathbb{C}\}$ is decreed to be a Y-weak open in X, as are all unions of finite intersections of such sets. By requiring Y to separate the points of X we guarantee that the Y-weak topology is Hausdorff, as follows. If $x_1 \neq x_2$ we may find φ in Y with $\varphi(x_1) \neq \varphi(x_2)$. Find disjoint open sets U_1 and U_2 in $\mathbb{C}$ containing, respectively, $\varphi(x_1)$ and $\varphi(x_2)$. The sets $\varphi^{-1}(U_1)$ and $\varphi^{-1}(U_2)$ are Y-weakly open sets in X, which are disjoint and contain x_1 and x_2, respectively.

At the moment, we are primarily interested in the following Y-weak topology. Let $X = \mathscr{A}^*$, the dual space of a Banach space $\mathscr{A}$, and let

$$Y = \{A^{**} \in \mathscr{A}^{**} : A \in \mathscr{A}\}.$$

Recall that for $A \in \mathscr{A}$, A^{**} is the bounded linear functional on $\mathscr{A}^*$ defined by $A^{**}(\ell) = \ell(A)$ for $\ell \in \mathscr{A}^*$. As a consequence of the Hahn–Banach theorem we know that $\|A^{**}\| = \|A\|$ (see the discussion following Corollary 3.5). We have $\{A^{**} : A \in \mathscr{A}\} \subseteq \mathscr{A}^{**}$, with this containment possibly proper. With the choice $X = \mathscr{A}^*, Y = \{A^{**} : A \in \mathscr{A}\}$ we call the Y-weak topology on $\mathscr{A}^*$ the *weak* topology* on $\mathscr{A}^*$. Summarizing, the weak* topology on the dual space $\mathscr{A}^*$ is the weakest topology allowing the linear functionals from $\mathscr{A}^*$ to $\mathbb{C}$ taking ℓ in $\mathscr{A}^*$ to $\ell(A) = A^{**}(\ell)$ to be continuous, for every $A \in \mathscr{A}$. Note that Y does separate the points of X: If $\ell_1 \neq \ell_2$ in $\mathscr{A}^*$, this means that there exists some A in $\mathscr{A}$ with $\ell_1(A) \neq \ell_2(A)$ and A^{**} is in Y with $A^{**}(\ell_1) \neq A^{**}(\ell_2)$. A sub-basis for the weak* topology is

$$\{(A^{**})^{-1}(U) : U \text{ is open in } \mathbb{C} \text{ and } A \in \mathscr{A}\}.$$

By taking finite intersections of these sub-basis elements we get a basis for the weak* topology. Every weak* open set is then a union of sets from this basis. The weak* open sets are also open in the norm topology, but not conversely; see Exercise 5.29.

Another weak topology on $X = \mathscr{A}^*$ of interest comes from the choice $Y = \mathscr{A}^{**}$. This is called the weak topology. When $\mathscr{A}$ is a Hilbert space, or more generally a reflexive Banach space, the weak and weak* topologies are the same, since $\mathscr{A}^{**} = \{A^{**} : A \in \mathscr{A}\}$ under the natural map. In general, the weak* topology is coarser than the weak topology (there are fewer open sets, and weak convergence implies weak* convergence).

In analysis we "understand" topologies by understanding convergence in the topology. As a first step in this direction, we describe when a *sequence* of points in $\mathscr{A}^*$ converges in the weak* topology. Then we will discuss a needed generalization of this result, with "sequence" replaced by "net," a concept defined later in this section.

Recall that a sequence $\{x_n\}$ in a topological space is said to converge to x if for every open set U containing x, there exists N (depending on U) so that $n \geq N$ implies $x_n \in U$. In a Hausdorff space, if a sequence converges its limit point must be unique.

Proposition 5.32. *Suppose $\{\varphi_n\}$ is a sequence in $\mathscr{A}^*$. We have $\varphi_n \to \varphi$ in the weak* topology if and only if $\varphi_n(A) \to \varphi(A)$ for each A in $\mathscr{A}$.*

Proof. If $\varphi_n \to \varphi$ (weak*) and A is in $\mathscr{A}$, then by definition A^{**} is continuous as a map of $(\mathscr{A}, \text{weak}^*)$ into $\mathbb{C}$, so $A^{**}(\varphi_n) \to A^{**}(\varphi)$. By the definition of A^{**}, this says $\varphi_n(A) \to \varphi(A)$, as desired.

Conversely, suppose $\varphi_n(A) \to \varphi(A)$ as $n \to \infty$ for each A in $\mathscr{A}$. We want to show that $\varphi_n \to \varphi$ (weak*). To this end, let $\mathscr{O}$ be a weak* open set containing φ. Our task is to show that there exists N so that if $n \geq N$, then φ_n is in $\mathscr{O}$. Now there is a basic open set containing φ and contained in $\mathscr{O}$; that is, there are points $A_1, A_2, \ldots, A_m$ in $\mathscr{A}$ and open sets $U_1, U_2, \ldots, U_m$ in $\mathbb{C}$ with

$$\varphi \in \bigcap_{j=1}^{m} (A_j^{**})^{-1}(U_j) \subseteq \mathscr{O}.$$

This means that $\varphi(A_j) = A_j^{**}(\varphi) \in U_j$ for $1 \leq j \leq m$. Since we are given that $\varphi_n(A) \to \varphi(A)$ as $n \to \infty$, for each A in $\mathscr{A}$, we have in particular, that for each j, $1 \leq j \leq m$, there is a finite N_j such that if $n \geq N_j$ then $\varphi_n(A_j) \in U_j$, or equivalently, $A_j^{**}(\varphi_n) \in U_j$. Let $N = \max\{N_j : 1 \leq j \leq m\}$. For $n \geq N$ we have

$$\varphi_n \in \bigcap_{j=1}^{m} (A_j^{**})^{-1}(U_j) \subseteq \mathscr{O}$$

and φ_n is in $\mathscr{O}$ for all $n \geq N$, as desired. $\square$

Proposition 5.32 explains why the weak* topology is often called the "topology of pointwise convergence."

Example 5.33. Consider the space ℓ^2, and recall that by Theorem 1.29, every bounded linear functional on ℓ^2 is given by $\langle \cdot, y \rangle$ for some $y \in \ell^2$. Let e_n be nth standard basis vector for ℓ^2, whose entries are all 0 except for a 1 in the nth position, and let φ_n denote the linear functional on ℓ^2 given by inner product with e_n: $\varphi_n(x) = \langle x, e_n \rangle$. We claim that $\varphi_n \to 0$ weak*. By Proposition 5.32, this will follow if we can show that

$$\varphi_n(x) \to 0$$

for each x in ℓ^2. By Parseval's identity,

$$\sum_{n=1}^{\infty} |\langle x, e_n \rangle|^2 < \infty$$

and hence $\langle x, e_n \rangle \to 0$ for each $x \in \ell^2$, as desired. Note that φ_n does not converge to 0 in the norm topology, since $\|\varphi_n\| = 1$.

In a metric space, sequences are adequate to detect limit points and continuity. The meaning of this assertion is as follows. Suppose X is a metric space and A is a subset of X. Let $\overline{A}$ denote the closure of A.

(a)   We have $x \in \overline{A}$ if and only if there is a *sequence* of points in A converging to x.
(b)   A function $f : X \to Y$, where Y is a topological space, is continuous if and only if whenever $\{x_n\}$ is a *sequence* in X with x_n converging to x, then the sequence $f(x_n)$ converges to $f(x)$.

The "only if" direction of (a) is not true in complete generality, that is, if X is a general topological space, it may be the case that a point in the closure of A cannot be reached as the limit of a sequence of points in A. An example is given in Exercise 5.30. Similarly the "if" direction of (b) may fail in some general topological spaces.

To recover results that resemble (a) and (b) but hold in general topological spaces we need to generalize the notion of a sequence to that of a "net."

Definition 5.34. A *directed set* is a set I together with a partial order $\leq$ such that if α and β are in I, then there exists γ in I with $\alpha \leq \gamma$ and $\beta \leq \gamma$.

We give some examples. If I is the set of positive integers $\mathbb{N}$, and we let $\leq$ be the usual ordering on $\mathbb{N}$, then $(\mathbb{N}, \leq)$ is a directed set. The collection I of all subsets of a given set S, partially ordered by inclusion, form a directed set. Two other useful examples of directed sets are as follows.

Example 5.35. In any topological space, let I be the collection of all open sets containing a fixed point x_0. Partially order the elements of I by reverse inclusion: $U \geq V$ means $U \subseteq V$ (bigger sets are smaller in the partial ordering). Given U_1 and U_2 in I, the open set $U_1 \cap U_2$ satisfies $U_1 \cap U_2 \geq U_1$ and $U_1 \cap U_2 \geq U_2$.

Example 5.36. Let S be any set and let I be all finite subsets of S. Partial order by inclusion, so $F_1 \geq F_2$ means $F_2 \subseteq F_1$. Given any pair of sets F_1 and F_2, $F_1 \cup F_2 \geq F_1$ and $F_1 \cup F_2 \geq F_2$.

One can think of a sequence in a set X as a function from $\mathbb{N}$ into X. The next definition generalizes this notion.

Definition 5.37. A *net* in a set X is a pair $((I, \leq), x)$ where $(I, \leq)$ is a directed set and x is a function from I into X. For α in I, denote $x(\alpha)$ by x_α. The net is denoted $\{x_\alpha\}_{\alpha \in I}$, and the set I is called the *index set* for the net. When X is a topological space, we say the net $\{x_\alpha\}_{\alpha \in I}$ converges to x_0 in X if for each open set U containing x_0 there exists $\alpha \in I$ such that $\beta \in I$ and $\beta \geq \alpha$ implies x_β is in U.

Nets are adequate to detect limit points. This is the content of the next result.

Theorem 5.38. *Suppose that A is a subset in a topological space X. We have $x \in \overline{A}$ if and only if there is a net of points in A converging to x.*

Proof. Suppose that x is in $\overline{A}$. Recall that $\overline{A}$ can be described as the set of points y such that every open set containing y intersects A. Let I be the collection of all open sets containing our given x, partially ordered by reverse inclusion; this is our directed set. For each $U \in I$, let x_U be a point of $A \cap U$. Thus $\{x_U\}_{U \in I}$ is a net and $x_U \to x$, since given any open set V containing x, if $U \geq V$ we have $x_U \in U \subseteq V$.

Conversely, if there is a net $\{x_\alpha\}$ of points in A converging to x, then any open set containing x will contain some points of this net, and thus some points of A. Therefore x is in $\overline{A}$. $\qquad\square$

Our next result shows that nets are adequate to detect continuity.

Theorem 5.39. *Suppose that X and Y are topological spaces. A function $f : X \to Y$ is continuous at x_0 in X if and only if the net $\{f(x_\alpha)\}_{\alpha \in I}$ converges to $f(x_0)$ in Y whenever $\{x_\alpha\}_{\alpha \in I}$ is a net in X converging to x_0.*

Proof. Assume that f is continuous at x_0 and suppose $\{x_\alpha\}$ is a net in X converging to x_0. Let V be open in Y with $f(x_0) \in V$. By continuity, $U \equiv f^{-1}(V)$ is open in X, and U contains x_0. There must exist β such that x_α is in U whenever $\alpha \geq \beta$. Thus for all $\alpha \geq \beta$, $f(x_\alpha)$ is in V; this says $f(x_\alpha) \to f(x_0)$. The converse direction is left as Exercise 5.31. $\qquad\square$

We can fully understand the weak* topology by understanding the convergence of nets. The next result is a generalization of Proposition 5.32, with sequences replaced by nets.

Theorem 5.40. *If $\{\varphi_\alpha\}_{\alpha \in I}$ is a net in $\mathscr{A}^*$ and φ is in $\mathscr{A}^*$, for some Banach space $\mathscr{A}$, then we have $\varphi_\alpha \to \varphi$ weak* in $\mathscr{A}^*$ if and only if $\varphi_\alpha(A) \to \varphi(A)$ for each A in $\mathscr{A}$.*

The proof, which is a straightforward modification of the proof of Proposition 5.32, is left to the reader as Exercise 5.27.

By the results in Section 4.1, and Exercise 4.3 in Chapter 4, we know that the norm-closed unit ball in a normed linear space X is compact (relative to the norm topology) if and only if X is finite-dimensional. Moving to infinite dimensions, one of the main virtues of using the weak* topology is to obtain compactness of the norm-closed unit ball in this new topology. The next result is traditionally called the Banach–Alaoglu theorem.

Theorem 5.41 (Banach–Alaoglu Theorem). *Let $\mathscr{A}^*$ be the dual space of some Banach space $\mathscr{A}$. The norm-closed unit ball, $\{\varphi \in \mathscr{A}^* : \|\varphi\| \leq 1\}$, is compact in the weak* topology.*

The proof of Theorem 5.41 depends on the Tychonoff theorem from topology, which says that an arbitrary product of compact sets is compact in the product topology. What is the product topology? If we have an indexed family of topological spaces $\{X_\alpha\}_{\alpha \in I}$, the product $\Pi_{\alpha \in I} X_\alpha$ is the set of all functions $f : I \to \cup_{\alpha \in I} X_\alpha$ such that $f(\alpha) \in X_\alpha$. When I is finite or countable, we think of $\Pi_\alpha X_\alpha$ as tuples or sequences $\{x_k\}$ where x_k is in X_k. Carrying over this notation to larger index sets we will write $\{x_\alpha\}_{\alpha \in I} \in \Pi_{\alpha \in I} X_\alpha$. For each $\beta \in I$ define

$$\pi_\beta : \Pi_{\alpha \in I} X_\alpha \to X_\beta$$

by $\pi_\beta(\{x_\alpha\}) = x_\beta$ or $\pi_\beta(f) = f(\beta)$, and call this the projection onto the βth factor. It assigns to each element of the product space its βth coordinate. The product topology on the product space is the weakest topology that makes each π_β continuous. Thus the sets

$$\{\pi_\beta^{-1}(U_\beta) : U_\beta \text{ is open in } X_\beta\},$$

where β is in I, are open in the product topology, and all finite intersections of sets of this type form a basis for the product topology. A net f_γ in $\Pi_{\alpha \in I} X_\alpha$ converges to f in the product topology if and only if

$$\pi_\beta(f_\gamma) \to \pi_\beta(f)$$

for all $\beta \in I$, or equivalently, if and only if $f_\gamma(\beta) \to f(\beta)$ for all β in I. To prove this statement, notice that one direction is immediate from the definition. The other direction follows similarly to the proof of Theorem 5.40. We are now ready to prove Theorem 5.41.

Proof (Theorem 5.41). Suppose that X^* is the dual of a Banach space X. Let B denote the norm-closed unit ball in X^*:

$$B = \{\varphi \in X^* : \|\varphi\| \le 1\}.$$

We wish to show that B is compact in the weak* topology.

For each x in X, consider the closed disk

$$D_x = \{\alpha \in \mathbb{C} : |\alpha| \le \|x\|\}$$

and note that D_x is compact in $\mathbb{C}$. By Tychonoff's theorem, the Cartesian product

$$C = \Pi_{x \in X} D_x$$

is compact in the product topology τ. Elements of C are functions f from the index set X to $\cup_{x \in X} D_x \subseteq \mathbb{C}$, assigning to each $x \in X$ a number $f(x)$ in D_x; that is, $|f(x)| \le \|x\|$ for each x in X. Each φ in B is a *linear* function from X to $\mathbb{C}$ of norm at most 1; that is $|\varphi(x)| \le \|x\|$ for each x in X. This means we can think of B as a subset of C, consisting of precisely those elements of C that are linear. Moreover, the weak* topology on B (inherited from the weak* topology on X^*) and the product topology on B considered as a subset of C coincide, since both are the topologies of pointwise convergence: A net $\{\varphi_\alpha\}$ in B converges weak* to φ if and only if $\varphi_\alpha(x) \to \varphi(x)$ for all $x \in X$, while $\varphi_\alpha \to \varphi$ in the product topology τ if and only if $\pi_x(\varphi_\alpha) \to \pi_x(\varphi)$ for each $x \in X$, or equivalently, if and only if $\varphi_\alpha(x) \to \varphi(x)$ for each $x \in X$.

We claim that B is closed in the product topology τ. If the claim is verified, then we see that B is a τ-closed subset of the τ-compact space C, and so it is τ-compact and hence also weak* compact. To verify the claim, suppose that φ_α is a net in B with $\varphi_\alpha \to f$ in the product topology. We want to show f is in B. We know that f

is in C, so that $|f(x)| \leq \|x\|$ for all x in X. Thus the only issue is whether f is linear. Let λ be in $\mathbb{C}$. We have $\varphi_\alpha(\lambda x) \to f(\lambda x)$ and $\varphi_\alpha(\lambda x) = \lambda \varphi_\alpha(x) \to \lambda f(x)$. This shows that $f(\lambda x) = \lambda f(x)$ for all $x \in X$ and scalars λ. A similar argument shows that $f(x_1 + x_2) = f(x_1) + f(x_2)$ and we conclude that f is linear.   □

Many names are justifiably associated with Theorem 5.41. As we have just seen, from the right perspective one could view it as a corollary to Tychonoff's theorem of 1930 (proved first for a product of compact *intervals* and extended by Čech in 1937 to more general spaces). In addition to Banach and Alaoglu, Bourbaki, Kakutani, and Shmulyan could all be mentioned for work during the period 1929–1939. Versions of the theorem in particular settings appeared earlier in work of Helly, Hilbert, and Riesz.

While the concept of a directed system or net comes from E. Moore and H. Smith, the terminology "net" is attributed to N. Steenrod. D. Sarason gives the following story about how it came into existence [42]:

> Kelley describes Steenrod as being especially creative in devising terminology. Steenrod is responsible for the term "net" as it is now commonly used for generalized sequence.... The term first appeared in this role in a 1950 paper of Kelley. Kelley had been planning to use the term "way"; that would have resulted in what we now call a "subnet" being referred to as a "subway." Steenrod, when informed by Kelley of his plan, apparently regarded Kelley's choice of terminology as frivolous, and after being prodded by Kelley, he suggested the term "net" as an alternative. His judgment prevailed (p. 18–19).

5.6 The Gelfand Transform

The ending point of the last section—the Banach–Alaoglu theorem—is our reentry point for continuing our studying of $\mathcal{M}_\mathscr{A}$, the maximal ideal space of $\mathscr{A}$, for a unital commutative Banach algebra $\mathscr{A}$. We now think of $\mathcal{M}_\mathscr{A}$ as sitting inside the norm-closed unit ball of $\mathscr{A}^*$ and let it inherit the weak* topology from $\mathscr{A}^*$. This is the only topology we will ever consider on $\mathcal{M}_\mathscr{A}$.

Theorem 5.42. *In the (relative) weak* topology, $\mathcal{M}_\mathscr{A}$ is a compact Hausdorff space.*

Proof. Since we already know that $\mathscr{A}^*$ with the weak* topology is Hausdorff, and a subspace of a Hausdorff space is Hausdorff, we need only check the compactness assertion. For this, it suffices to show $\mathcal{M}_\mathscr{A}$ is closed in the weak* topology; then the Banach–Alaoglu theorem will guarantee that, as a closed subset of the weak* compact set $\{\ell \in \mathscr{A}^* : \|\ell\| \leq 1\}$, $\mathcal{M}_\mathscr{A}$ is compact.

To see that $\mathcal{M}_\mathscr{A}$ is weak* closed, let $\{\varphi_\alpha\}_{\alpha \in I}$ be a net in $\mathcal{M}_\mathscr{A}$ converging to $\varphi \in \mathscr{A}^*$, that is, $\varphi_\alpha \to \varphi$ weak *. To show φ is in $\mathcal{M}_\mathscr{A}$ we show it is multiplicative. Let A and B be in $\mathscr{A}$. Since $\varphi_\alpha \to \varphi$ weak*, we have $\varphi_\alpha(A) \to \varphi(A)$, $\varphi_\alpha(B) \to \varphi(B)$, and $\varphi_\alpha(AB) \to \varphi(AB)$. On the one hand,

$$\varphi_\alpha(AB) = \varphi_\alpha(A)\varphi_\alpha(B) \to \varphi(A)\varphi(B),$$

while on the other hand $\varphi_\alpha(AB) \to \varphi(AB)$. In a Hausdorff space, a net converges to at most one point (see Exercise 5.31), and thus $\varphi(AB) = \varphi(A)\varphi(B)$, and φ is multiplicative. Since $1 = \varphi_\alpha(I) \to \varphi(I)$, φ is nontrivial, and thus φ is in $\mathcal{M}_\mathscr{A}$. $\quad\square$

Since $X = \mathcal{M}_\mathscr{A}$ is a compact Hausdorff space, $C(X) = C(\mathcal{M}_\mathscr{A})$, the space of continuous (complex-valued) functions on $\mathcal{M}_\mathscr{A}$ in the supremum norm, is a unital Banach algebra. There is a natural candidate for a linear map from $\mathscr{A}$ to $C(\mathcal{M}_\mathscr{A})$, namely the map that sends $A \in \mathscr{A}$ to the function whose value at φ is $\varphi(A)$. Explicitly, let

$$\Gamma : \mathscr{A} \to C(\mathcal{M}_\mathscr{A})$$

be defined by $\Gamma(A)(\varphi) = \varphi(A)$ for φ in $\mathcal{M}_\mathscr{A}$. We do need to verify that $\Gamma(A)$ is *continuous* on $\mathcal{M}_\mathscr{A}$, and we do this momentarily. More briefly, we will write $\hat{A}$ for $\Gamma(A)$, so that $\hat{A}$ will be the (continuous) function on $\mathcal{M}_\mathscr{A}$ taking the value $\varphi(A)$ at φ. We will call $\hat{A}$ the Gelfand transform of A, and the map Γ itself the Gelfand map, or Gelfand transform, of $\mathscr{A}$.

Theorem 5.43. *Let $\mathscr{A}$ be a commutative unital Banach algebra. For each A in $\mathscr{A}$, $\hat{A}$ is continuous on $\mathcal{M}_\mathscr{A}$. The map Γ is a continuous homomorphism of the commutative unital Banach algebra $\mathscr{A}$ into $C(\mathcal{M}_\mathscr{A})$. Furthermore,*

$$\|\Gamma(A)\|_\infty = r(A) \le \|A\|$$

for all A, and Γ has norm 1. The Gelfand map is one-to-one if and only if the intersection of all of the maximal ideals of $\mathscr{A}$ is $\{0\}$.

Proof. The continuity of $\hat{A}$ follows immediately from the definition of the weak* topology: If φ_α is a net in $\mathcal{M}_\mathscr{A}$ converging weak* to φ, then

$$\hat{A}(\varphi_\alpha) \equiv \varphi_\alpha(A) \to \varphi(A) = \hat{A}(\varphi).$$

By Theorem 5.39, this shows that $\hat{A}$ is continuous.

It is easy to check that Γ is linear and preserves multiplication; we leave the details to the reader. Recalling that $\mathcal{M}_\mathscr{A}$ is compact,

$$\|\hat{A}\|_\infty = \max\{|\hat{A}(\varphi)| : \varphi \in \mathcal{M}_\mathscr{A}\} = \max\{|\varphi(A)| : \varphi \in \mathcal{M}_\mathscr{A}\} = r(A),$$

since $\sigma(A) = \{\varphi(A) : \varphi \in \mathcal{M}_\mathscr{A}\}$. Since $r(A) \le \|A\|$, this says $\|\Gamma(A)\|_\infty \le \|A\|$ and the Gelfand map is contractive, hence also continuous. Consideration of the identity shows that the Gelfand map has norm 1.

Since $\Gamma(A)$ is the zero function if and only if $\varphi(A) = 0$ for all $\varphi \in \mathcal{M}_\mathscr{A}$, that is, if and only if A lies in the kernel of every $\varphi \in \mathcal{M}_\mathscr{A}$, the kernel of the map Γ is the intersection of all the maximal ideals of $\mathscr{A}$. This gives the final statement. $\quad\square$

The intersection of all maximal ideals of $\mathscr{A}$ is called the *radical* of $\mathscr{A}$. When the radical of $\mathscr{A}$ is $\{0\}$, $\mathscr{A}$ is called *semisimple*. Thus Theorem 5.43 says that in a commutative unital Banach algebra $\mathscr{A}$, the Gelfand map is one-to-one if and only if $\mathscr{A}$ is semisimple.

There is more that can be said when $\mathscr{A}$ is a unital, commutative C^*-algebra. Informally, the next result says that every such C^*-algebra "is" $C(X)$ for some compact Hausdorff space X.

Theorem 5.44. *If $\mathscr{A}$ is a unital, commutative C^*-algebra, the Gelfand map Γ is an isometric $*$-isomorphism of $\mathscr{A}$ onto $C(\mathscr{M}_{\mathscr{A}})$.*

Proof. By virtue of Theorem 5.43, we only need to check that Γ is $*$-preserving, isometric, and bijective. We'll start with the $*$-preserving property, i.e., we show that $\Gamma(A^*) = \overline{\Gamma(A)}$, or, in our briefer notation, $\widehat{A^*} = \overline{\hat{A}}$ for arbitrary A. Given $\varphi \in \mathscr{M}_{\mathscr{A}}$, $\widehat{A^*}(\varphi) = \varphi(A^*)$ and $\hat{A}(\varphi) = \overline{\varphi(A)}$. Recall (Theorem 5.21) that a complex homomorphism of a C^*-algebra is automatically a $*$- homomorphism, so $\varphi(A^*) = \overline{\varphi(A)}$. This verifies $\widehat{A^*} = \overline{\hat{A}}$.

We know from Theorem 5.43 that $\|\hat{A}\|_\infty \leq \|A\|$ for all $A \in \mathscr{A}$. If in addition A is self-adjoint, then

$$\|A\| = r(A) = \|\hat{A}\|_\infty = \|\Gamma(A)\|_\infty.$$

In particular, for *any* A we have $\|\Gamma(A^*A)\|_\infty = \|A^*A\|$, since A^*A is self-adjoint. Thus for any A

$$\|\Gamma(A)\|_\infty^2 = \|\Gamma(A)^*\Gamma(A)\|_\infty = \|\Gamma(A^*A)\|_\infty = \|A^*A\| = \|A\|^2,$$

where we have used the C^*-identity in the first and last equalities, and the fact that Γ is a $*$-homomorphism. This line of computation shows that Γ is an isometry, hence it is a $*$-isomorphism onto its image in $C(\mathscr{M}_{\mathscr{A}})$.

To finish the proof, we need to show that the image of $\mathscr{A}$ under Γ is all of $C(\mathscr{M}_{\mathscr{A}})$. To do this, we use the Stone–Weierstrass theorem (see Section A.4 in the Appendix). We claim that $\Gamma(\mathscr{A})$ is a closed subspace of $C(\mathscr{M}_{\mathscr{A}})$ which is also a subalgebra of $C(\mathscr{M}_{\mathscr{A}})$, containing the constant functions, separating the points of $\mathscr{M}_{\mathscr{A}}$, and closed under conjugation. Since Γ is isometric and a $*$-homomorphism, we see that $\Gamma(\mathscr{A})$ is a closed subalgebra which is also closed under conjugation. Since $\Gamma(\lambda I)$ is the constant function λ, this subalgebra contains the constants. Its clear that it separates points, since if $\varphi_1 \neq \varphi_2$ in $\mathscr{M}_{\mathscr{A}}$, then we may find $A \in \mathscr{A}$ with $\varphi_1(A) \neq \varphi_2(A)$ and thus $\Gamma(A)(\varphi_1) \neq \Gamma(A)(\varphi_2)$. Invoking the Stone–Weierstrass theorem we conclude that $\Gamma(\mathscr{A}) = C(\mathscr{M}_{\mathscr{A}})$. $\qquad\square$

Example 5.45. Start with any compact Hausdorff space X and let $\mathscr{A} = C(X)$ be the commutative unital C^*-algebra of continuous functions on X, in the supremum norm as usual. Notice that X is homeomorphic to $\mathscr{M}_{\mathscr{A}}$ in the weak* topology, via the map that sends $x \in X$ to the multiplicative linear functional ev_x of evaluation at x; to see that this mapping is onto $\mathscr{M}_{\mathscr{A}}$ we are making use of Example 5.27 in Section 5.3. What is the Gelfand map $\Gamma : C(X) \to C(\mathscr{M}_{\mathscr{A}})$? By definition, we have $\Gamma(f) = \hat{f}$ where $\hat{f}(\varphi) = \varphi(f)$, for $f \in \mathscr{A} = C(X)$ and $\varphi \in \mathscr{M}_{\mathscr{A}}$. But each φ has the form ev_x for some $x \in X$, and $\hat{f}(ev_x) = ev_x(f) = f(x)$. Thus "$\Gamma(f) = f$" once we identify $x \in X$ with ev_x in $\mathscr{M}_{\mathscr{A}}$.

We seem to be severely limited by the "commutativity" hypothesis of the last theorem, but we will next see one important situation in which we can get around this. We again focus on a unital C^*-algebra $\mathscr{A}$. For any subset S of $\mathscr{A}$, let $C^*(S)$ denote the C^*-algebra generated by the elements of S together with the identity I; this is the intersection of all C^*-algebras containing I and S. For the case that the set S is the singleton $\{A\}$, we simply write $C^*(A)$. This is the norm-closure of all finite linear combinations of "words" in A, A^*, I, where a "word" is a finite product $T_1 T_2 \cdots T_n$ with $T_j \in \{A, A^*, I\}$ for each j. In the special case that A is normal, then we may permute A and A^* and thus $C^*(A)$ is the closure of all polynomials in A and A^*, $p(A, A^*) = \sum c(m,n) A^n (A^*)^m$, where n, m are nonnegative integers.

Clearly, if A is normal, then $C^*(A)$ is a commutative, unital C^*-algebra, and by the Gelfand theory, $C^*(A)$ is isometrically $*$-isomorphic to the C^*-algebra of continuous functions on the maximal ideal space of $C^*(A)$ via the Gelfand transform Γ. What we would like to be able to do is identify this maximal ideal space more explicitly, that is, we seek to replace $\mathscr{M}_{C^*(A)}$ by a set homeomorphic to it via a "natural" map. Moreover, this identification should depend on A in some clear way. This is the content of our next result. As motivation, recall that in any commutative Banach algebra, $\lambda \in \sigma(A)$ if and only if $\lambda = \varphi(A)$ for some φ in the maximal ideal space. In the situation we're currently interested in, where our commutative Banach algebra is $C^*(A)$ for some normal A, φ is unique: $\varphi_1(A) = \varphi_2(A)$ implies that $\varphi_1(A^*) = \varphi_2(A^*)$, since $\varphi_j(A^*) = \overline{\varphi_j(A)}$ for $j = 1, 2$, and thus φ_1 agrees with φ_2 on any polynomial $p(A, A^*)$ by linearity and multiplicativity, and hence on all of $C^*(A)$, by continuity. This suggests we may be able to identify the maximal ideal space of $C^*(A)$ with $\sigma(A)$, the spectrum of A, via a homeomorphism. Such a homeomorphism would then allow us to produce a $*$-isomorphism between $C^*(A)$ and $C(\sigma(A))$.

Theorem 5.46. *Suppose $\mathscr{A}$ is a singly generated, commutative, unital C^*-algebra, with $\mathscr{A} = C^*(A)$ for some A which is necessarily normal. There is a unique $*$-isomorphism of $\mathscr{A}$ onto $C(\sigma(A))$ mapping A to the identity function on $\sigma(A)$.*

Proof. We know that $\mathscr{A}$ is the closure of the polynomials in A and A^*, and that the Gelfand map Γ is an isometric $*$-isomorphism of $C^*(A)$ onto $C(\mathscr{M}_{\mathscr{A}})$. Define $\tau : \mathscr{M}_{\mathscr{A}} \mapsto \sigma(A)$ by $\tau(\varphi) = \varphi(A)$ (of course, τ is just $\hat{A}$, but we find it convenient to use this alternate notation). We are using the fact that

$$\sigma(A) = \{\varphi(A) : \varphi \in \mathscr{M}_{\mathscr{A}}\};$$

this depends on the commutativity of $\mathscr{A}$. Thus τ maps onto $\sigma(A)$ and is continuous (where we use the weak* topology on $\mathscr{M}_{\mathscr{A}}$). The map τ is also one-to-one; this is the uniqueness observation we made before the statement of the theorem. Thus τ, being a continuous bijection from the compact set $\mathscr{M}_{\mathscr{A}}$ to the Hausdorff space $\sigma(A)$, is a homeomorphism. The homeomorphism τ induces a $*$-isomorphism $h : C(\mathscr{M}_{\mathscr{A}}) \mapsto C(\sigma(A))$ by $h(f) = f\tau^{-1}$ (the reader is invited to check the details of this statement in Exercise 5.33). Thus $h \circ \Gamma$ is a $*$-isomorphism from $C^*(A)$ onto $C(\sigma(A))$, and under this isomorphism A gets mapped to $\hat{A} \circ \tau^{-1} \in C(\sigma(A))$. What is $\hat{A} \circ \tau^{-1}$? We claim this is just the identity function on $\sigma(A)$: For $\lambda \in \sigma(A)$, $\tau^{-1}(\lambda) = \varphi$ if and only if $\lambda = \tau(\varphi)$ if and only if $\varphi(A) = \lambda$, and thus $\hat{A}(\tau^{-1}(\lambda)) = \hat{A}(\varphi) = \varphi(A) = \lambda$, where φ is the unique multiplicative linear functional with $\varphi(A) = \lambda$.

Finally we claim that this is the *unique* *-isomorphism of $C^*(A)$ onto $C(\sigma(A))$ sending A to the identity function $\zeta(z) = z$. If A is mapped to ζ, then A^* is mapped to $\overline{\zeta}$ and any polynomial $p(A,A^*)$ in A and A^* must be mapped to the corresponding polynomial $p(z,\overline{z})$ in z and $\overline{z}$. The uniqueness statement then follows, since $\mathscr{A}$ is the closure of the polynomials $p(A,A^*)$. $\square$

Recall that if Banach algebras $\mathscr{A}$ and $\mathscr{B}$ have a common identity and $A \in \mathscr{B} \subseteq \mathscr{A}$, then by Exercise 5.18, $\sigma_{\mathscr{A}}(A) \subseteq \sigma_{\mathscr{B}}(A)$ where the containment may be strict, while $\partial\sigma_{\mathscr{B}}(A) \subseteq \partial\sigma_{\mathscr{A}}(A)$. We will see that if $\mathscr{A}$ and $\mathscr{B}$ are required to be C^*-algebras, we have $\sigma_{\mathscr{A}}(A) = \sigma_{\mathscr{B}}(A)$, a result we will refer to as "spectral permanence." To prove this, we need a lemma. It reduces the issue of invertibility to the question of invertibility of self-adjoint elements.

Lemma 5.47. *Suppose $\mathscr{A}$ is a unital C^*-algebra. An element $A \in \mathscr{A}$ is invertible in $\mathscr{A}$ if and only if AA^* and A^*A are both invertible.*

Proof. The "only if" direction is trivial, so we omit its proof. Suppose now that there exists $B \in \mathscr{A}$ with $B(A^*A) = I$, so that BA^* is a left inverse for A. Similarly, if $C \in \mathscr{A}$ satisfies $(AA^*)C = I$, then A^*C is a right inverse for A. Having both a right and left inverse, A is invertible. $\square$

Theorem 5.48. *Suppose $\mathscr{A}$ and $\mathscr{B}$ are C^*-algebras with common identity I, and assume $\mathscr{B} \subseteq \mathscr{A}$. If $A \in \mathscr{B}$, then $\sigma_{\mathscr{A}}(A) = \sigma_{\mathscr{B}}(A)$.*

Proof. First suppose A is self-adjoint, and let $\mathscr{C} = C^*(A)$, the C^*-algebra generated by $A = A^*$ and I. Since $\mathscr{C}$ is commutative and unital, the spectrum of the self-adjoint element A in $\mathscr{C}$ is real (Theorem 5.31), and thus $\sigma_{\mathscr{C}}(A) = \partial\sigma_{\mathscr{C}}(A)$. Now

$$\sigma_{\mathscr{B}}(A) \subseteq \sigma_{\mathscr{C}}(A) = \partial\sigma_{\mathscr{C}}(A) \subseteq \partial\sigma_{\mathscr{B}}(A) \subseteq \sigma_{\mathscr{B}}(A),$$

so we must have equality throughout and $\sigma_{\mathscr{C}}(A) = \sigma_{\mathscr{B}}(A)$ for self-adjoint A. Similarly,

$$\sigma_{\mathscr{A}}(A) \subseteq \sigma_{\mathscr{C}}(A) = \partial\sigma_{\mathscr{C}}(A) \subseteq \partial\sigma_{\mathscr{A}}(A) \subseteq \sigma_{\mathscr{A}}(A),$$

and $\sigma_{\mathscr{C}}(A) = \sigma_{\mathscr{A}}(A)$. Thus we have shown that for A self-adjoint, $\sigma_{\mathscr{A}}(A) = \sigma_{\mathscr{B}}(A)$.
For arbitrary $A \in \mathscr{B}$ we use Lemma 5.47. We know that

$$\sigma_{\mathscr{A}}(A) \subseteq \sigma_{\mathscr{B}}(A)$$

so we need only show the reverse inclusion. To this end, let $\lambda \in \sigma_{\mathscr{B}}(A)$, so that $A - \lambda I$ is not invertible in $\mathscr{B}$, and by Lemma 5.47, either $(A - \lambda I)^*(A - \lambda I)$ or $(A - \lambda I)(A - \lambda I)^*$ is not invertible in $\mathscr{B}$. Since both are self-adjoint, either $(A - \lambda I)^*(A - \lambda I)$ or $(A - \lambda I)(A - \lambda I)^*$ is not invertible in $\mathscr{A}$ by the first part of the proof. Applying Lemma 5.47 once more, we see that λ is in $\sigma_{\mathscr{A}}(A)$ and we have established the inclusion we needed, completing the proof. $\square$

We can give an important application of the last theorem. Recall that a self-adjoint element of a commutative unital C^* algebra has *real* spectrum. Now suppose that A is self-adjoint in the unital, but not necessarily commutative, C^*-algebra $\mathscr{A}$. Since $C^*(A)$ is a commutative C^*-subalgebra of $\mathscr{A}$ containing the identity of $\mathscr{A}$ (by definition), Theorem 5.48 asserts

$$\sigma_{\mathscr{A}}(A) = \sigma_{C^*(A)}(A) \subseteq \mathbb{R}.$$

In fact, more is true, as the next result shows.

Theorem 5.49. *If $\mathscr{A}$ is a unital C^*-algebra, and $A \in \mathscr{A}$ is normal, then*

(a)   *A is self-adjoint if and only if $\sigma(A) \subseteq \mathbb{R}$.*
(b)   *A is unitary if and only if $\sigma(A) \subseteq \partial\mathbb{D}$.*
(c)   *$A^2 = A$ if and only if $\sigma(A) \subseteq \{0, 1\}$.*

Proof. We give the proof of (a) and leave the remaining parts to the reader as Exercise 5.35.

Let $\mathscr{B}$ be the commutative C^*-algebra $C^*(A)$; Theorem 5.48 tells us that $\sigma_{\mathscr{B}}(A) = \sigma_{\mathscr{A}}(A)$. The Gelfand map $\Gamma : \mathscr{B} \to C(\mathcal{M}_{\mathscr{B}})$ is a $*$-isomorphism, so A is self-adjoint if and only if $\hat{A}$ is self-adjoint as an element of $C(\mathcal{M}_{\mathscr{B}})$, i.e., if and only if $\hat{A}$ is real-valued. By the commutativity of $\mathscr{B}$, the range of $\hat{A}$ is $\{\varphi(A) : \varphi \in \mathcal{M}_{\mathscr{B}}\} = \sigma_{\mathscr{B}}(A)$. We conclude that A is self-adjoint if and only if $\sigma_{\mathscr{B}}(A)$, or equivalently $\sigma_{\mathscr{A}}(A)$, is contained in $\mathbb{R}$. $\square$

An even briefer proof of the last result goes as follows: By Theorem 5.46 there is a unique $*$-isomorphism of $C^*(A)$ onto $C(\sigma(A))$ which takes A to the identity function on $\sigma(A)$. Note Theorem 5.48 says we need not specify whether we mean the spectrum of A relative to $C^*(A)$, or relative to some larger C^*-algebra $\mathscr{A}$ containing $C^*(A)$. Then A is self-adjoint if and only if the identity function is real-valued on $\sigma(A)$, that is, if and only if $\sigma(A) \subseteq \mathbb{R}$.

Theorems 5.48 and 5.46 provide us with the necessary tools to describe the *continuous functional calculus*, and we will turn to this in the next section.

We close this section with an example of historical importance.

Example 5.50. The space $\ell^1(\mathbb{Z})$ of doubly infinite sequences $\{a_n\}_{n=-\infty}^{\infty}$ is a Banach space in the norm $\|\{a_n\}\| \equiv \sum_{n=-\infty}^{\infty} |a_n|$. In this example we will consider a Banach space, which we will denote W, which is isometrically isomorphic to $\ell^1(\mathbb{Z})$, and see that there is multiplication on W that makes it a Banach algebra. The space W consists of all complex-valued functions f on the unit circle T that can be expressed as an absolutely convergent Fourier series:

$$f(e^{i\theta}) = \sum_{n=-\infty}^{\infty} a_n e^{in\theta} \tag{5.2}$$

normed by

$$\|f\| = \|f\|_W \equiv \sum_{n=-\infty}^{\infty} |a_n| < \infty.$$

The series defining f converges uniformly on T, so that any $f \in W$ is continuous on T. The mapping that sends $f \in W$ to its sequence of Fourier coefficients $\{\hat{f}(n)\}_{n=-\infty}^{\infty}$

is clearly a linear isometry mapping W onto $\ell^1(\mathbb{Z})$. If we define multiplication of functions in W to be pointwise multiplication, W becomes a Banach algebra. To see this, suppose that f is as in Equation (5.2) and that g is also in W, with

$$g(e^{i\theta}) = \sum_{n=-\infty}^{\infty} b_n e^{in\theta}.$$

Since both series converge absolutely,

$$(fg)(e^{i\theta}) = f(e^{i\theta})g(e^{i\theta}) = \sum_{n=-\infty}^{\infty} \left(\sum_{k=-\infty}^{\infty} a_k b_{n-k} \right) e^{in\theta}$$

so that

$$\|fg\| = \sum_{n=-\infty}^{\infty} \left| \sum_{k=-\infty}^{\infty} a_k b_{n-k} \right|$$

$$\leq \sum_{n=-\infty}^{\infty} \sum_{k=-\infty}^{\infty} |a_n||b_{n-k}|$$

$$= \left(\sum_{n=-\infty}^{\infty} |a_n| \right) \left(\sum_{j=-\infty}^{\infty} |b_j| \right)$$

$$= \|f\|\|g\|.$$

Clearly W is commutative, and the constant function 1 serves as an identity, so W, called the *Wiener algebra*, is a commutative unital Banach algebra.

What is the maximal ideal space $\mathcal{M}_W$? Each $\lambda \in T$ determines a multiplicative linear functional φ_λ defined by evaluation at λ; that is $\varphi_\lambda(f) = f(\lambda)$. Notice that

$$|\varphi_\lambda(f)| = |f(\lambda)| = \left| \sum_{n=-\infty}^{\infty} a_n \lambda^n \right| \leq \sum_{n=-\infty}^{\infty} |a_n| = \|f\|$$

and $\varphi(1) = 1$, so that $\|\varphi_\lambda\| = 1$ as expected. We claim that

$$\mathcal{M}_W = \{\varphi_\lambda : \lambda \in T\}.$$

Suppose that φ is an arbitrary multiplicative linear functional in $\mathcal{M}_W$ and let $\lambda = \varphi(\chi)$ where $\chi(e^{i\theta}) = e^{i\theta}$. We have

$$|\lambda| = |\varphi(\chi)| \leq \|\varphi\|\|\chi\| = (1)(1) = 1.$$

Also,

$$1 = \varphi(1) = \varphi(\chi\chi^{-1}) = \varphi(\chi)\varphi(\chi^{-1})$$

so that

$$\varphi(\chi^{-1}) = 1/\varphi(\chi) = 1/\lambda.$$

Since

$$|1/\lambda| = |\varphi(\chi^{-1})| \leq \|\varphi\| \|\chi^{-1}\| = 1$$

we must have $|\lambda| = 1$. To complete the verification of the claim, we show that $\varphi = \varphi_\lambda$, evaluation at our point λ in the circle. Given

$$f(e^{i\theta}) = \sum_{n=-\infty}^{\infty} a_n e^{in\theta},$$

we set

$$f_N(e^{i\theta}) = \sum_{n=-N}^{N} a_n e^{in\theta}$$

so that

$$\|f_N - f\| = \sum_{|n|=N+1}^{\infty} |a_n| \to 0$$

as $N \to \infty$. By the continuity of φ,

$$\begin{aligned}
\varphi(f) &= \lim_{N \to \infty} \varphi \left(\sum_{n=-N}^{N} a_n \chi^n \right) \\
&= \lim_{N \to \infty} \sum_{n=-N}^{N} a_n \varphi(\chi^n) \\
&= \lim_{N \to \infty} \sum_{n=-N}^{N} a_n \varphi(\chi)^n \\
&= \sum_{n=-\infty}^{\infty} a_n \lambda^n \\
&= \varphi_\lambda(f),
\end{aligned}$$

as desired. It follows that the map $\Psi : T \to \mathcal{M}_W$ defined by $\Psi(\lambda) = \varphi_\lambda$ carries T onto $\mathcal{M}_W$, and it is clearly one-to-one. If we give T the usual (metric) topology and $\mathcal{M}_W$ the weak* topology, this mapping Ψ is continuous. To verify this, note that if $\lambda_n \to \lambda$ in T,

$$\Psi(\lambda_n)(f) = f(\lambda_n) \to f(\lambda) = \Psi(\lambda)(f)$$

for every $f \in W$, and $\Psi(\lambda_n)$ tends to $\Psi(\lambda)$ in the weak* topology. Since both T and $\mathcal{M}_W$ are compact and Hausdorff, we see that Ψ is a homeomorphism of T onto $\mathcal{M}_W$.

We can now easily describe the Gelfand transform $\Gamma : W \to C(\mathcal{M}_W)$:

$$\Gamma(f)(\varphi_\lambda) = \varphi_\lambda(f) = f(\lambda).$$

Though we will stick with correct notation, as in the previous line, the reader may find it useful to also think of our homeomorphism Ψ as an "identity map," identifying φ_λ with λ and $\mathcal{M}_W$ with T (as we did in Example 5.45). From this point of view, $\Gamma(f)$ is f. In other words, Γ maps f, considered as an element of W, to f considered as an element of $C(T)$.

The proof we give of the next result, called Wiener's theorem, is often considered to be the first spectacular success of Gelfand theory. Wiener's original proof of this result relied on a difficult classical analysis argument. By contrast, the elegant proof provided later by Gelfand uses only elementary Banach algebra theory.

Theorem 5.51. *Suppose $f(e^{i\theta}) = \sum_{-\infty}^{\infty} a_n e^{in\theta}$ lies in W. If f does not vanish on T, then $1/f$ is also in W, that is, there exist $\{b_n\}$ with $\sum_{-\infty}^{\infty} |b_n| < \infty$ and*

$$\frac{1}{f(e^{i\theta})} = \sum_{n=-\infty}^{\infty} b_n e^{in\theta}.$$

The statement of the theorem can also be phrased as follows: If a function f in $C(T)$ has an absolutely convergent Fourier series and is nonzero on T, then $1/f$ has an absolutely convergent Fourier series.

Proof (Theorem 5.51). The hypothesis on f says that $\varphi_\lambda(f) = f(\lambda)$ does not vanish as λ ranges over T. Since we have shown that the functionals φ_λ exhaust $\mathscr{M}_W$, we may apply Corollary 5.29 to conclude that f is invertible in W, which is the desired conclusion.   □

In Exercise 5.37 you are asked to show that the Gelfand transform on the Wiener algebra is not isometric, and as a consequence W cannot be made into a C^*-algebra.

5.7 The Continuous Functional Calculus

Let's recap where we are from the last section: Given a normal element N in a unital C^*-algebra $\mathscr{A}$, the C^*-algebra $C^*(N)$ generated by N, N^*, and I, which is the closure of the polynomials in N and N^*, is a commutative C^*-algebra. Spectral permanence says

$$\sigma_{\mathscr{A}}(A) = \sigma_{C^*(N)}(A)$$

for any $A \in C^*(N)$, so the subscript can be omitted without danger of misinterpretation. Theorem 5.46 says there is a unique isometric $*$-isomorphism γ from $C^*(N)$ onto $C(\sigma(N))$ which pairs N with the identity function z on $\sigma(N)$. Under this $*$-isomorphism, N^* and $\bar{z}$ are paired, as are cI and the constant function c, and a polynomial $p(N, N^*)$ is paired with the corresponding polynomial $p(z, \bar{z})$. Now if f is any continuous function on $\sigma(N)$, we *define* $f(N)$ to be the corresponding element $\gamma^{-1}(f)$ of $C^*(N)$; note this is in agreement with our previous observations when f is a polynomial in z and $\bar{z}$. Clearly if $f, g \in C(\sigma(N))$, so that $f + g$ and fg are also, then

$$(f+g)(N) = f(N) + g(N)$$

and

$$(fg)(N) = f(N)g(N),$$

since γ^{-1} is linear and multiplicative. This assignment $f \mapsto f(N)$ is called *the continuous functional calculus* or the *functional calculus for normal elements*. It is illustrated schematically below.

$$C^*(N) \underset{\gamma^{-1}}{\overset{\gamma}{\rightleftarrows}} C(\sigma(N))$$

$$I \longleftrightarrow 1$$

$$N \longleftrightarrow z$$

$$N^* \longleftrightarrow \bar{z}$$

$$N^j N^{*k} \longleftrightarrow z^j \bar{z}^k$$

$$p(N, N^*) \longleftrightarrow p(z, \bar{z})$$

$$f(N) \longleftrightarrow f$$

Making this definition of $f(N)$ immediately raises the issue of how it behaves with respect to spectra. The next result—the full version of the spectral mapping theorem—is to be compared with Theorem 5.14. As observed above, there is no need to distinguish between $\sigma_{\mathscr{A}}$ and $\sigma_{C^*(N)}$ in this result.

Theorem 5.52. *Suppose N is a normal element in a unital C^*-algebra $\mathscr{A}$ and let $f \in C(\sigma(N))$. We have $\sigma(f(N)) = f(\sigma(N))$.*

Proof. Since $f \mapsto f(N) \equiv \gamma^{-1}(f)$ is a $*$-isomorphism of $C(\sigma(N))$ onto $C^*(N)$ we have

$$\sigma(f(N)) = \sigma_{C(\sigma(N))}(f) = \text{range } f = f(\sigma(N)).$$

$\square$

As an application, recall that we previously observed (as a consequence of the spectral radius formula) that for a self-adjoint element A in a unital C^*-algebra, $r(A) = \|A\|$. We can now obtain a deeper result, namely that the equality of the norm and spectral radius holds for any *normal* element. If N is normal in a unital C^*-algebra, then by Theorem 5.46 there is an isometric $*$-isomorphism of $C^*(N)$ with $C(\sigma(N))$ which pairs N with z. Thus

$$\|N\| = \|z\|_{\infty, C(\sigma(N))} = \max\{|z| : z \in \sigma(N)\} = r(N).$$

Thus we have shown the following result.

Theorem 5.53. *For any normal element N of a unital C^*-algebra, $\|N\| = r(N)$.*

Definition 5.54. A self-adjoint element A of a unital C^*-algebra $\mathscr{A}$ is said to be *positive* if its (necessarily real) spectrum is contained in $[0, \infty)$; we write $A \geq 0$ in this case. The positive elements of $\mathscr{A}$ are denoted $\mathscr{A}_+$.

For example, in the C^*-algebra $C(X)$, where X is a compact Hausdorff space, the positive elements are precisely the nonnegative functions.

The functional calculus for normal operators plays a role in the next result on roots of positive elements.

Theorem 5.55. *If $A \in \mathscr{A}_+$ and $n \in \mathbb{N}$, there is a unique $B \in \mathscr{A}_+$ satisfying $B^n = A$.*

Proof. We establish existence first. Since by assumption, $\sigma(A) \subseteq [0,\infty)$, the real-valued function $f(t) = t^{1/n}$ is continuous on $\sigma(A)$. Thus $f(A)$ is defined by the functional calculus; set $B = f(A)$. Is B self-adjoint? Yes, since f is real-valued, hence self-adjoint as an element of $C(\sigma(A))$. Moreover, by Theorem 5.52,

$$\sigma(B) = \sigma(f(A)) = f(\sigma(A)) \subseteq [0,\infty)$$

so that B is positive. Finally, $B^n = f^n(A) = A$, since f^n is the identity function.

For the uniqueness statement, we first note that uniqueness is clear if $\mathscr{A} = C(X)$ for some compact Hausdorff space X; this is simply the statement that a nonnegative number has a *unique* nonnegative nth root. Now suppose, in a general setting, that B and C are positive nth roots of the positive element A, with $B = f(A)$ for $f(t) = t^{1/n}$. Note that C commutes with A since $CA = CC^n = C^n C = AC$, hence C commutes with any polynomial in A. Now by the Weierstrass approximation theorem, $f(t) = t^{1/n}$ is a uniform limit of a sequence of polynomials p_n, on say $[0, \|A\|]$. Since $B = f(A) = \lim_{n\to\infty} p_n(A)$, where each $p_n(A)$ commutes with C, we can conclude that B commutes with C.

Consider $\mathscr{B} = C^*(B,C)$, the smallest unital C^*-algebra containing the self-adjoint elements B and C. It contains $A = B^n$, and by the above argument is easily seen to be commutative, so that the Gelfand transform Γ is an isometric $*$-isomorphism of $\mathscr{B}$ onto $C(\mathscr{M}_\mathscr{B})$. Since A, B, and C are positive elements of $\mathscr{B}$, $\hat{A}, \hat{B}$, and $\hat{C}$ are positive elements of $C(\mathscr{M}_\mathscr{B})$, i.e., they are nonnegative functions. We have

$$(\hat{B})^n = \hat{A} = \widehat{C^n} = (\hat{C})^n,$$

so that by our observations on uniqueness in $C(\mathscr{M}_\mathscr{B})$ we must have $\hat{B} = \hat{C}$, and hence $B = C$ as desired. $\square$

We can use the definition of positive element to give an ordering on the self-adjoint elements of a unital C^*-algebra.

Definition 5.56. *If A and B are self-adjoint elements of a unital C^*-algebra we say $A \leq B$ if $B - A$ is positive.*

Two important facts about positive elements in a unital C^*-algebra $\mathscr{A}$ are as follows:

- If $A \in \mathscr{A}$, then $A^*A \geq 0$.
- If $A \in \mathscr{A}$ and $B \in \mathscr{A}$ are positive, then so is $A + B$.

The converse of the first property, that any positive element of an arbitrary unital C^*-algebra $\mathscr{A}$ is of the form A^*A for some $A \in \mathscr{A}$ is a consequence of the $n = 2$ case of Theorem 5.55. The second property shows that "$\leq$" is a transitive relation on the self-adjoint elements of a unital C^*-algebra. It, together with the obvious result that $A \in \mathscr{A}_+$ and $t \geq 0$ implies $tA \in \mathscr{A}_+$, say that $\mathscr{A}_+$ forms a "cone" in $\mathscr{A}$. While we will not prove either of the two properties above in general, we will verify them in the particular context of positive elements of $\mathscr{B}(\mathscr{H})$.

Proposition 5.57. *Let $\mathcal{H}$ be a Hilbert space. We have the following:*

(a) *If $T \in \mathcal{B}(\mathcal{H})$ satisfies $\langle Th, h \rangle \geq 0$ for all $h \in \mathcal{H}$, then T is positive, i.e., a positive element of the C^*-algebra $\mathcal{B}(\mathcal{H})$.*
(b) *For any $T \in \mathcal{B}(\mathcal{H})$, the operator T^*T is positive.*
(c) *The sum of two positive operators in $\mathcal{B}(\mathcal{H})$ is positive.*

Proof. In part (a), the hypothesis that $\langle Th, h \rangle \geq 0$ for all $h \in \mathcal{H}$ implies that T is self-adjoint (see Exercise 4.12 in Chapter 4), and so its spectrum is contained in $\mathbb{R}$. We want to show further that its spectrum is contained in $[0, \infty)$. Let $t < 0$ and consider, for arbitrary $h \in \mathcal{H}$,

$$|\langle (T - tI)h, h \rangle| = |\langle Th, h \rangle - t\|h\|^2| = \langle Th, h \rangle + |t|\|h\|^2 \geq |t|\|h\|^2.$$

Since

$$|t|\|h\|^2 \leq |\langle (T - tI)h, h \rangle| \leq \|(T - tI)h\|\, \|h\|$$

we must have

$$\|(T - tI)h\| \geq |t|\|h\|$$

and thus the operator $T - tI$ is bounded below. This says that t is not an approximate eigenvalue of T. Since the spectrum of T is contained in the real line, and every point of the boundary of the spectrum is an approximate eigenvalue (see Exercise 5.5), this tells us that t is not in the spectrum of T.

Part (b) follows immediately from (a) since $\langle T^*Th, h \rangle = \langle Th, Th \rangle \geq 0$ for all $h \in \mathcal{H}$.

Finally, suppose that A and B are positive operators in $\mathcal{B}(\mathcal{H})$. Each has a positive square root $A^{1/2}$ and $B^{1/2}$, by Theorem 5.55. For any $h \in \mathcal{H}$ we have

$$\begin{aligned}
\langle (A + B)h, h \rangle &= \langle Ah, h \rangle + \langle Bh, h \rangle \\
&= \langle A^{1/2}h, A^{1/2}h \rangle + \langle B^{1/2}h, B^{1/2}h \rangle \\
&= \|A^{1/2}h\|^2 + \|B^{1/2}h\|^2 \geq 0
\end{aligned}$$

and by (a) we conclude that $A + B$ is positive, as desired.   □

Further properties of "$\leq$" are given in Exercise 5.45, including some that indicate potential pitfalls of the notation.

5.8 Fredholm Operators

The invertible elements in $\mathcal{B}(\mathcal{H})$ are those operators which are one-to-one and onto the Hilbert space $\mathcal{H}$. Here we will consider a related class of operators obtained by relaxing these conditions.

Definition 5.58. A bounded linear operator T on a Hilbert space $\mathcal{H}$ is called *Fredholm* if

(1)   the range of T is closed,

(2)   the dimension of the kernel of T is finite, and

(3)   the dimension of the quotient $\mathcal{H}/T\mathcal{H}$ is finite.

When T has closed range, notice that by Exercise 3.26 of Chapter 3, the dimension of $\mathcal{H}/T\mathcal{H}$ is equal to the dimension of $(\operatorname{ran} T)^{\perp} \equiv (T\mathcal{H})^{\perp}$.

Every invertible operator is clearly Fredholm. A simple example of a noninvertible operator that is Fredholm is given by the shift S on ℓ^2, where we have $\dim(\ker S) = 0$ and $\dim(\ell^2/S\ell^2) = 1$. When T is compact and $\lambda \neq 0$, then $T - \lambda I$ is Fredholm; this follows from Exercise 4.10 in Chapter 4 and the results of Section 4.6. This can be rephrased by saying that a compact perturbation of the identity is a Fredholm operator, where the terminology "compact perturbation of the identity" refers to an operator of the form $I - T$, where T is compact.

Although our interest in these operators will be confined to the Hilbert space setting, the definition of "Fredholm operator" can equally well be made in the context of a bounded linear operator on a Banach space X (or even from one Banach space to another): $T \in \mathcal{B}(X)$ is Fredholm if the range of T is closed, the dimension of the kernel of T is finite, and the dimension of X/TX is finite. There is some redundancy in this definition; T automatically has closed range if X/TX is finite-dimensional (here we are considering the quotient X/TX simply as a vector space, and the assumption is that this vector space has a finite Hamel basis); this was the content of Exercise 3.33 in Chapter 3.

The next result, known as Atkinson's theorem, characterizes the Fredholm operators on $\mathcal{H}$ as those operators which are "invertible modulo the compacts," a concept we introduced at the end of Section 5.3.

Theorem 5.59 (Atkinson's Theorem). *Suppose T is in $\mathcal{B}(\mathcal{H})$ for some Hilbert space $\mathcal{H}$. The operator T is Fredholm if and only if there is a bounded operator S in $\mathcal{B}(\mathcal{H})$ such that $ST - I$ and $TS - I$ are both compact.*

Proof. First suppose that T is Fredholm, so that the kernel of T is finite-dimensional, as is $(T\mathcal{H})^{\perp}$. The restriction of T to the closed subspace $(\ker T)^{\perp}$ is a one-to-one map of $(\ker T)^{\perp}$ onto $T\mathcal{H}$. Thus we may define S on $T\mathcal{H}$ to be the inverse of this restriction:

$$S|_{T\mathcal{H}} = \left(T|_{(\ker T)^{\perp}}\right)^{-1}.$$

Define S to be 0 on $(T\mathcal{H})^{\perp}$. On $(\ker T)^{\perp}$ we have $ST - I = 0$, and hence the dimension of the range of $ST - I$ is finite. On $T\mathcal{H}$, $TS - I = 0$, and thus the dimension of the range of $TS - I$ is finite. This shows that $ST - I$ and $TS - I$ are finite rank operators, and hence compact.

Conversely, suppose that there exists $S \in \mathcal{B}(\mathcal{H})$ with $ST - I$ and $TS - I$ both compact. We have

$$(ST - I)|_{\ker T} = -I|_{\ker T}.$$

Since $ST - I$ is compact, this forces the kernel of T to be finite-dimensional. Taking the adjoint of $TS - I$, we see that on the kernel of T^*, $S^*T^* - I$ agrees with $-I$, and since $S^*T^* - I$ is compact, this forces the kernel of T^* to be finite-dimensional.

The kernel of T^* is equal to $(\operatorname{ran} T)^\perp$ (Exercise 2.16 in Chapter 2). Thus we can conclude that T is Fredholm as soon as we have shown that T has closed range, since in this case the dimension of $\mathcal{H}/T\mathcal{H}$ is equal to the dimension of $(\operatorname{ran} T)^\perp$.

Since $ST - I$ is compact, Theorem 4.29 tells us that ST has closed range. By Exercise 3.27 of Chapter 3, it follows that ST is bounded below on $(\ker ST)^\perp$, and thus there exists $c > 0$ such that

$$\|STh\| \geq c\|h\|$$

for all $h \in (\ker ST)^\perp$. From this it follows that

$$\|Th\| \geq \frac{c}{\|S\|}\|h\|$$

for all $h \in (\ker ST)^\perp$. The reader can now easily show, using completeness, that

$$\{Th : h \in (\ker ST)^\perp\}$$

is a closed subspace of $\mathcal{H}$.

Now write K for the compact operator $ST - I$. Then $\ker(ST) = \ker(I + K)$, where $\ker(I + K)$ is finite-dimensional by Exercise 4.10 in Chapter 4, and so $T(\ker ST)$ is a finite-dimensional subspace of $\mathcal{H}$. Thus we have a finite-dimensional subspace

$$\{Th : h \in \ker(ST)\},$$

and a closed subspace

$$\{Th : h \in \ker(ST)^\perp\}.$$

We may apply Exercise 4.4 in Chapter 4 to conclude that

$$\{Th : h \in \ker(ST)\} + \{Th : h \in \ker(ST)^\perp\}$$

is a closed subspace. Since this is obviously the range of T, we are done. $\qquad\square$

We can rephrase Theorem 5.59 as "$T \in \mathcal{B}(\mathcal{H})$ is Fredholm if and only if $T + \mathcal{K}$ is invertible in the Calkin algebra $\mathcal{B}(\mathcal{H})/\mathcal{K}(\mathcal{H})$." As an immediate consequence we see that the product of two Fredholm operators is Fredholm. This result has a purely algebraic proof as well; see [41].

The terminology "Fredholm operator" recognizes the pioneering work of Erik Fredholm. In 1903 he published a paper that, in modern language, dealt with equations of the form

$$f(s) - \int_a^b k(s,t)f(t)dt = g(s), \tag{5.3}$$

where $k(s,t)$ is in $L^2([a,b] \times [a,b])$ and f, g are in $L^2[a,b]$. A natural question to ask is: For which g does a solution f exist to this equation, and when a solution exists for a particular g, can the solutions be described? From our modern perspective we can think of Equation (5.3) as

$$(I - K)f = g,$$

where K is the integral operator with kernel $k(s,t)$ as defined in Section 2.1. Furthermore, as we know from Theorem 4.16, the operator K is compact (in fact, Hilbert–Schmidt). The extent to which solutions to Equation (5.3) are not unique is measured by the dimension of the kernel of $I - K$, and the extent to which solutions (with g being given and f being the unknown) fail to exist is measured by the dimension of $[\mathrm{ran}\,(I - K)]^\perp$. Notice how Theorem 4.32 elaborates on this: If the dimension of the kernel of $I - K$ is zero, then $I - K$ is invertible and for each g a unique solution f exists.

5.9 Exercises

5.1. Show that a Banach algebra $\mathscr{A}$ with an involution satisfying

$$\|A^*A\| \geq \|A\|^2$$

is a C^*-algebra, meaning that equality holds in this inequality.

5.2. Suppose that $\mathscr{A}$ is a C*-algebra.

(a) Show that if $\mathscr{A}$ has a unit, it is unique (call it I); furthermore $I^* = I$ and $\|I\| = 1$ (provided $\|A\| \neq 0$ for some $A \in \mathscr{A}$).
(b) Suppose $\mathscr{A}$ is unital. Show that if A is invertible, so is A^*, with $(A^*)^{-1} = (A^{-1})^*$.
(c) Every $A \in \mathscr{A}$ can be written as $A = X + iY$ where X and Y are self-adjoint.
(d) If $\mathscr{A}$ is unital and U is unitary (meaning $UU^* = U^*U = I$), then $\|U\| = 1$.

5.3. Let $\mathscr{G}$ denote the set of invertible elements in a unital Banach algebra. Show that the map of $\mathscr{G}$ into $\mathscr{G}$ defined by $A \to A^{-1}$ is continuous.

5.4. Suppose that $F : \Omega \to \mathscr{A}$ is a function defined on an open set $\Omega \subseteq \mathbb{C}$ and taking values in a Banach space $\mathscr{A}$. Show that if f is strongly analytic in Ω, then it is weakly analytic (as defined in Section 5.2).

5.5. Recall that for $T \in \mathscr{B}(\mathscr{H})$, the operator $T - \lambda I$ is invertible if and only if $T - \lambda I$ is bounded below and has dense range. So one way for λ to get into the spectrum of T is for $T - \lambda I$ to *not* be bounded below, meaning that there are unit vectors h_n with $\|(T - \lambda I)h_n\| \to 0$. A point λ with this property is said to be an *approximate eigenvalue* of T; the set of all approximate eigenvalues of T is called the *approximate point spectrum* of T. Show

(a) Every eigenvalue of T is in the approximate point spectrum of T.
(b) The approximate point spectrum of T is a closed set (show its complement is open).
(c) Show that if T_n is invertible for all n and $T_n \to T$ where T is not invertible, then 0 is an approximate eigenvalue of T. Hints: Explain why it suffices to show that if the range of T is not dense, then there are unit vectors h_n with $\|Th_n\| \to 0$. Then assume that the range of T is not dense and find a nonzero vector h with $h \perp \overline{\mathrm{ran}\,T}$ (why must such an h exist?). Consider $h_n = T_n^{-1}h/\|T_n^{-1}h\|$.

(d) If λ is in the boundary of $\sigma(T)$, then show that λ is an approximate eigenvalue for T.

(e) Extend the result of (d) to the case that T is a bounded linear operator on a Banach space X, with "approximate eigenvalue" defined in the analogous way.

5.6. Suppose that $T \in \mathcal{B}(\mathcal{H})$. Show that λ is not an approximate eigenvalue of T if and only if $T - \lambda I$ has a left inverse.

5.7. Let $\sigma_{ap}(A)$ denote the approximate point spectrum for an operator $A \in \mathcal{B}(\mathcal{H})$.

(a) Show that $\Pi_{j=1}^{n}(A - \lambda_j I)$ is bounded below on $\mathcal{H}$ if and only if $A - \lambda_j I$ is bounded below for $1 \leq j \leq n$.

(b) Show that for any polynomial p, $\sigma_{ap}(p(A)) = p(\sigma_{ap}(A))$.

Does the analogous result, with "approximate point spectrum" replaced by "point spectrum" hold? The point spectrum of A is $\{\lambda : \ker(A - \lambda) \text{ is nontrivial}\}$, i.e., the set of eigenvalues of A.

5.8. Suppose that A is a bounded linear operator on a Hilbert space $\mathcal{H}$. Show that if $A - \lambda I$ does not have dense range in $\mathcal{H}$, then $\overline{\lambda}$ is an eigenvalue of A^*, and conversely, if μ is an eigenvalue of A^*, then $A - \overline{\mu}I$ does not have dense range. Thus the compression spectrum of A can be described in terms of the eigenvalues of A^*.

5.9. (An Inversion Spectral Mapping Theorem.) Suppose that A is an invertible operator in $\mathcal{B}(\mathcal{H})$. The goal of this problem is to show that

$$\sigma(A^{-1}) = \left\{ \frac{1}{\lambda} : \lambda \in \sigma(A) \right\}.$$

(a) Show that if $A - \lambda I$ is not bounded below, then $A^{-1} - \frac{1}{\lambda}I$ is not bounded below, and conversely that if $A^{-1} - \mu I$ is not bounded below, then $A - \frac{1}{\mu}I$ is not bounded below. Show that the eigenvalues of A^{-1} are precisely the reciprocals of the eigenvalues of A.

(b) Show that $A - \lambda I$ fails to have dense range in $\mathcal{H}$ if and only if $A^{-1} - \frac{1}{\lambda}I$ fails to have dense range in $\mathcal{H}$. Exercise 5.8 may be helpful here.

Conclude that

$$\sigma(A^{-1}) = \left\{ \frac{1}{\lambda} : \lambda \in \sigma(A) \right\}.$$

This result holds more generally for any invertible element in a unital Banach algebra; see for example, p. 204 in [8].

5.10. Consider the operator on ℓ^∞ defined by

$$T(x_1, x_2, \ldots) = (\lambda_1 x_1, \lambda_2 x_2, \ldots)$$

where $(\lambda_1, \lambda_2, \ldots)$ is in ℓ^∞. Find $\sigma_p(T)$, $\sigma(T)$ and show that $\sigma(T) \backslash \sigma_p(T)$ is the residual spectrum of T.

5.11. Recall from Exercise 2.10 in Chapter 2 that if W is a weighted shift and $\lambda \in \mathbb{C}$ satisfies $|\lambda| = 1$, then λW is a weighted shift which is unitarily equivalent to W. Show that weighted shifts have "circularly symmetric" spectra, that is, if $\mu \in \sigma(W)$ and $|\lambda| = 1$, then $\lambda\mu \in \sigma(W)$.

5.12. Recall the Banach space

$$C^1[0,1] = \{f : f \text{ is continuously differentiable on } [0,1]\}$$

with norm $\|f\|_\infty + \|f'\|_\infty$.

(a) Show that under pointwise multiplication, $C^1[0,1]$ is a Banach algebra. Is it a C^*-algebra if we define $f^* = \bar{f}$?
(b) Let $g(x) = x$ for $x \in [0,1]$. What is the norm of g in $C^1[0,1]$? What is the spectral radius $r(g)$?
(c) Show that for each closed set $E \subseteq [0,1]$,

$$\mathscr{J}_E \equiv \{f \in C^1[0,1] : f(x) = 0 \text{ for } x \in E\}$$

is a closed, two-sided ideal in $C^1[0,1]$.
(d) Find a closed ideal in $C^1[0,1]$ which is not of the form $\mathscr{J}_E$ as in (c).

5.13. Suppose $\mathscr{A}$ is a Banach algebra and $\mathscr{J}$ is a proper, closed ideal. Show that

$$(A + \mathscr{J})(B + \mathscr{J}) = AB + \mathscr{J}$$

is a well-defined multiplication on $\mathscr{A}/\mathscr{J}$ under which this quotient space becomes a complex algebra.

5.14. Recall that an operator $T \in \mathscr{B}(X)$, where X is a Banach space, is an isometry if $\|Tx\| = \|x\|$ for all $x \in X$.

(a) Show that the spectrum of an isometry T is contained in the unit circle $\partial\mathbb{D}$ if T is invertible.
(b) Show that if T is an isometry but is not invertible, then its spectrum is $\overline{\mathbb{D}}$. Hint: By Exercise 5.5, the boundary of the spectrum is contained in the set of approximate eigenvalues of T.
(c) Give an example of a continuous $\varphi : [0,1] \to [0,1]$ so that the composition operator C_φ (see Exercise 2.3 in Chapter 2) is an isometry on $C[0,1]$ and $\sigma(C_\varphi) = \overline{\mathbb{D}}$.

5.15. (a) An operator $T \in \mathscr{B}(\mathscr{H})$ is said to be *nilpotent* if $T^n = 0$ for some positive integer n. Show any nilpotent operator has spectrum equal to $\{0\}$.
(b) Say T is *quasinilpotent* if $\sigma(T) = \{0\}$. By (a), every nilpotent operator is quasinilpotent. Show that the operator $T : \ell^2 \to \ell^2$ given by

$$T(x_1, x_2, \dots) = (0, \frac{x_1}{2}, \frac{x_2}{4}, \dots, \frac{x_n}{2^n}, \dots)$$

is quasinilpotent.

5.16. Consider the Volterra integral operator V acting on $L^2([0,1], dx)$ defined by

$$Vf(x) = \int_0^x f(t)dt.$$

(a) Show that for any positive integer n,

$$V^{n+1} f(x) = \frac{1}{n!} \int_0^x (x-t)^n f(t)dt.$$

(b) Show that $\sigma(V) = \{0\}$.

5.17. Let $\mathscr{A}$ be the Banach algebra $C(T)$ in the supremum norm, where T denotes the unit circle $\partial\mathbb{D}$. Let $\mathscr{B}$ be the subalgebra of $C(T)$ consisting of those $f \in C(T)$ for which there exist polynomials p_n in z with p_n converging uniformly to f on T.

(a) Show that $g(z) = \bar{z}$ is not in $\mathscr{B}$ (but of course it is in $\mathscr{A}$).
(b) Consider the function $f(z) = z$ which is in both $\mathscr{A}$ and $\mathscr{B}$. What is $\sigma_{\mathscr{A}}(f)$? Show that $\sigma_{\mathscr{B}}(f) = \bar{\mathbb{D}}$, the closed unit disk. Observe that although $\sigma_{\mathscr{A}}(f) \neq \sigma_{\mathscr{B}}(f)$, the spectral radius of the element f doesn't change in passing from $\mathscr{A}$ to $\mathscr{B}$.

5.18. Suppose $\mathscr{A}$ and $\mathscr{B}$ are Banach algebras with common identity and $\mathscr{B} \subseteq \mathscr{A}$. Show $\sigma_{\mathscr{A}}(A) \subseteq \sigma_{\mathscr{B}}(A)$ and $\partial\sigma_{\mathscr{B}}(A) \subseteq \partial\sigma_{\mathscr{A}}(A)$, for any $A \in \mathscr{B}$. Hint for the second part: Since the first part implies that the interior of $\sigma_{\mathscr{A}}(A)$ is contained in the interior of $\sigma_{\mathscr{B}}(A)$ for any A in $\mathscr{B}$, argue first that it suffices to show that if $\lambda \in \partial\sigma_{\mathscr{B}}(A)$, then $\lambda \in \sigma_{\mathscr{A}}(A)$.

5.19. Suppose that $\mathscr{A}$ is a C^*-algebra with unit $I_{\mathscr{A}}$, $\mathscr{B}$ is a C^*-algebra with unit $I_{\mathscr{B}}$ and $\rho : \mathscr{A} \to \mathscr{B}$ is a $*$-homomorphism with $\rho(I_{\mathscr{A}}) = I_{\mathscr{B}}$. Prove the following:

(a) For every $A \in \mathscr{A}$, $\sigma(\rho(A)) \subseteq \sigma(A)$, and hence $r(\rho(A)) \leq r(A)$.
(b) For every $A \in \mathscr{A}$, $\|\rho(A)\| \leq \|A\|$.
(c) If ρ is a $*$-isomorphism, then ρ is an isometry.

5.20. Find the norm of the operator on $\mathscr{B}(\mathscr{H})$ where $\mathscr{H} = \mathbb{C}^2$ which is given by the matrix

$$\begin{bmatrix} a & b \\ c & d \end{bmatrix}.$$

Give your answer in terms of the numbers $S = |a|^2 + |b|^2 + |c|^2 + |d|^2$ and $D = ad - bc$.

5.21. Suppose that $\|\cdot\|_1$ and $\|\cdot\|_2$ are two norms on a $*$-algebra $\mathscr{A}$, each of which make $\mathscr{A}$ into a C^*-algebra. Show $\|\cdot\|_1 = \|\cdot\|_2$.

5.22. Show that the noncommutative unital Banach algebra $\mathbb{M}_n(\mathbb{C})$ of all $n \times n$ matrices with complex entries has no nontrivial two-sided ideals. (Hints: Take any nonzero matrix A. Show that by multiplying A on the left and right by the appropriate sequence of matrices you can isolate any entry of A and move it anywhere you want. Recall that the elementary row and column operations of interchanging two rows or two columns of A can be obtained by multiplying A by the appropriate elementary matrix.)

5.23. Suppose that $\mathcal{J}$ is a closed two-sided ideal in $\mathcal{B}(\mathcal{H})$, for $\mathcal{H}$ a Hilbert space. The goal of this problem is to show that either $\mathcal{J} = \{0\}$, or $\mathcal{J}$ contains $\mathcal{K}(\mathcal{H})$, the ideal of compact operators on $\mathcal{H}$. (Compare this with the statement in Exercise 5.22.)

(a) Suppose T is a nonzero operator in $\mathcal{J}$. Find vectors f_0, f_1 with $Tf_0 = f_1$ and $f_1 \neq 0$. Show that if g_0, g_1 are any pair of nonzero vectors in $\mathcal{H}$, then the rank one operator S defined by

$$Sf = \frac{\langle f, g_0 \rangle g_1}{\|g_0\|^2}$$

is in $\mathcal{J}$. By Exercise 2.6 in Chapter 2, this will show that $\mathcal{J}$ contains all rank 1 operators. Hint: Let

$$Af = \frac{\langle f, g_0 \rangle f_0}{\|g_0\|^2}$$

and

$$Bf = \frac{\langle f, f_1 \rangle g_1}{\|f_1\|^2}$$

and compute BTA.
(b) Apply Exercise 4.9 of Chapter 4 to show that $\mathcal{J}$ contains all finite rank operators.

5.24. Let (X, Ω, μ) be a σ-finite measure space and suppose $\varphi \in L^\infty(\mu)$. Define M_φ on $L^2(\mu)$ by $M_\varphi(f) = \varphi f$, so that M_φ is the multiplication operator with symbol φ. Recall that M_φ is a bounded linear operator on $L^2(\mu)$ with $\|M_\varphi\| = \|\varphi\|_\infty$.

(a) Show that M_φ is normal, with $M_\varphi^* = M_{\overline{\varphi}}$.
(b) Show that $\varphi \to M_\varphi$ is a $*$- homomorphism from $L^\infty(\mu)$ into $\mathcal{B}(L^2(\mu))$.
(c) Show that the eigenvalues of M_φ are the complex numbers λ for which $\varphi^{-1}(\{\lambda\})$ has positive measure, and that $\sigma(M_\varphi)$ is the essential range of φ. The essential range of φ is defined as:

$$\{w \in \mathbb{C} : \mu\{x : |f(x) - w| < \varepsilon\} > 0 \text{ for all } \varepsilon > 0\}.$$

(d) Show directly that any closed set in $\mathbb{C}$ that contains the range of φ must contain the essential range of φ.
(e) Suppose f is a continuous function on $\sigma(M_\varphi)$. Identify $f(M_\varphi)$ in the continuous functional calculus. (Hint: Make a guess, and use the uniqueness statement for the functional calculus to prove your guess correct.)

5.25. Suppose that A is a self-adjoint operator in $\mathcal{B}(\mathcal{H})$ for some Hilbert space $\mathcal{H}$. Show that if $\ker(A - tI) = \{0\}$ and $A - tI$ has closed range for some real number t, then the range of $A - tI$ is $\mathcal{H}$. Conclude that a self-adjoint operator has no residual spectrum.

5.26. Suppose S is a set and τ_1, τ_2 are topologies on S with τ_1 weaker than τ_2. For an arbitrary set A in S, how does the closure of A relative to τ_1 compare to the closure of A relative to τ_2? Is it easier for a set to be compact in the τ_1-topology or the τ_2-topology? Is it easier for a sequence (or net) to converge in the τ_1-topology or the τ_2-topology?

5.27. Prove Theorem 5.40 by modifying the proof of Proposition 5.32.

5.28. This problem explains why we require Y to be a vector space in defining the Y-weak topology.

(a) Suppose that X is a vector space and $\varphi_1, \varphi_2, \ldots, \varphi_n$ are linear maps from X into $\mathbb{C}$. Let φ be a linear map from X into $\mathbb{C}$. Show that φ is in the linear span of $\{\varphi_1, \varphi_2, \ldots, \varphi_n\}$ if and only if

$$\ker \varphi_1 \cap \ker \varphi_2 \cap \cdots \cap \ker \varphi_n \subseteq \ker \varphi.$$

(b) Suppose that Y is a vector space of linear functionals on X that separates the points of X. Show that a linear functional φ on X is continuous with respect to the Y-weak topology if and only if φ is in Y.

5.29. Suppose $\mathscr{A}$ is an infinite-dimensional Banach space and $\mathscr{A}^*$ is its dual space.

(a) Show that a neighborhood basis of 0 in the weak* topology is given by the collection of sets

$$\mathcal{O}_{A_1, \ldots, A_n} \equiv \{\varphi \in \mathscr{A}^* : |\varphi(A_j)| < 1, 1 \leq j \leq n\},$$

where n is a positive integer, and $A_1, A_2, \ldots, A_n$ are in $\mathscr{A}$. To do this, you need to show that $\mathcal{O}_{A_1, \ldots, A_n}$ is a weak* open set containing 0, and for any weak* open set $\mathcal{O}$ containing 0, there is some positive integer n, and points $A_j \in \mathscr{A}$ with $0 \in \mathcal{O}_{A_1, \ldots, A_n} \subseteq \mathcal{O}$.

(b) Let φ_0 be in $\mathscr{A}^*$. In the weak* topology, a sub-basic neighborhood of φ_0 has the form

$$\{\varphi \in \mathscr{A}^* : |\varphi(A) - \varphi_0(A)| < \varepsilon\},$$

where $\varepsilon > 0$ and A is fixed in $\mathscr{A}$. Basic neighborhoods of φ_0 are finite intersections of sub-basic neighborhoods:

$$N = \{\varphi \in \mathscr{A}^* : |\varphi(A_j) - \varphi_0(A_j)| < \varepsilon_j, 1 \leq j \leq n\}$$

where $A_j \in \mathscr{A}$. Show that these are always unbounded sets. Hint: Look at a sub-basic neighborhood and first suppose $\varphi_0(A) = 0$, so that the sub-basic neighborhood contains $\{\varphi : \varphi(A) = 0\}$, a subspace of $\mathscr{A}^*$.

(c) Show that the open unit ball in the norm of $\mathscr{A}^*$ is not weak* open.

5.30. Let $X = [0, 1]$ and put a topology on X by declaring the open sets to be the empty set and those subsets of X whose complement is at most countable. Consider the set $A = [0, 1)$. Show that the closure of A is $[0, 1]$, so that in particular 1 lies in the closure of A. Show that there is no sequence $\{x_n\}$ of points in $[0, 1)$ that converges to 1.

5.31.(a) Suppose S is a Hausdorff topological space. Show that a net in S converges to at most one point; i.e., if $x_\alpha \to x$ and $x_\alpha \to y$ then $x = y$.
(b) Give a proof of the "if" direction of Theorem 5.39.

5.32. Let $\mathscr{H}$ be a Hilbert space with orthonormal basis $\{e_n\}_0^\infty$ and consider the set $E = \{e_m + me_n : 0 \leq m < n, n = 1, 2, 3, \ldots\}$. Note that E is countable. Show that 0 is in the weak closure of E, but there is no *sequence* $x_n \in E$ with $x_n \to 0$ weakly.

5.33. Prove the following statement used in Theorem 5.46: If X and Y are homeomorphic compact Hausdorff spaces, then $C(X)$ and $C(Y)$ are $*$-isomorphic unital C^*-algebras in a natural way.

5.34. Let $\mathscr{A}$ be a unital commutative Banach algebra, and suppose $A, B \in \mathscr{A}$. Show that $r(A + B) \leq r(A) + r(B)$ and $r(AB) \leq r(A)r(B)$. (Hint: Use the Gelfand transform.) Show the same result holds if $\mathscr{A}$ is not assumed to be commutative, provided $AB = BA$. Show the result fails in general (look in $\mathbb{M}_2(\mathbb{C})$).

5.35. Suppose $\mathscr{A}$ is a unital C^*-algebra, and $A \in \mathscr{A}$ is a normal element. Show that A is unitary if and only if $\sigma(A) \subseteq \partial\mathbb{D}$, the unit circle in the complex plane. Show that $A^2 = A$ if and only if $\sigma(A) \subseteq \{0, 1\}$.

5.36. Suppose that N is a normal operator in $\mathscr{B}(\mathscr{H})$ for some Hilbert space $\mathscr{H}$. If λ is in $\mathbb{C} \backslash \sigma(N)$, show that

$$\|(N - \lambda I)^{-1}\| = \text{dist}\, (\sigma(N), \lambda)^{-1}$$

where dist $(\sigma(N), \lambda)$ is the distance from $\sigma(N)$ to λ.

5.37. Let W denote the Wiener algebra. Show that the Gelfand transform $\Gamma : W \to C(\mathscr{M}_W)$ is not isometric, and indeed is not even bounded below. Thus W cannot be made into a C^*-algebra (for example, by defining $f^* = \bar{f}$).

5.38. The "one-sided Wiener algebra" W_+ is defined to be the set of all f in the Wiener algebra W of the form

$$f(e^{i\theta}) = \sum_{n=0}^{\infty} a_n e^{in\theta}.$$

(a) Show that W_+ is a closed subalgebra of W.
(b) For λ in the closed unit disk $\bar{\mathbb{D}}$, show that φ_λ taking $\sum_{n=0}^{\infty} a_n e^{in\theta} \in W_+$ to $\sum_{n=0}^{\infty} a_n \lambda^n$ is a multiplicative linear functional on W_+.
(c) Show that

$$\mathscr{M}_{W_+} = \{\varphi_\lambda : \lambda \in \bar{\mathbb{D}}\}$$

and the map $\lambda \to \varphi_\lambda$ is a homeomorphism of $\bar{\mathbb{D}}$ and $\mathscr{M}_{W_+}$, the latter being equipped with the weak* topology.
(d) If $f \in W_+$, describe the spectra $\sigma_+(f)$ and $\sigma(f)$ of f as an element of W_+ and W, respectively.

5.39. Consider the Banach space $\ell^1(\mathbb{N}_0)$ of sequences $\{x_n\}_{n=0}^\infty$ in the norm

$$\|\{x_n\}\| = \sum_{n=0}^\infty |x_n|.$$

For $\{x_n\}$ and $\{y_n\}$ in $\ell^1(\mathbb{N}_0)$, define the convolution $\{x_n\} * \{y_n\}$ to be the sequence $\{z_n\}$ defined by

$$z_n = \sum_{k=0}^n x_k y_{n-k}$$

for $n = 0, 1, 2, \ldots$.

(a) Show that $\ell^1(\mathbb{N}_0)$ becomes a commutative unital Banach algebra under the convolution product.

(b) Show that as a Banach algebra, $\ell^1(\mathbb{N}_0)$ is isometrically isomorphic to the one-sided Wiener algebra W_+ defined in Exercise 5.38.

(c) For $0 < a < 1$, set $x = \{a^n\} = (1, a, a^2, a^3, \ldots)$. Find $\sigma(x)$, the spectrum of x in $\ell^1(\mathbb{N}_0)$.

5.40. Consider the Banach space c of convergent sequences, as defined in Exercise 3.20 of Chapter 3. Define a product and involution respectively on c by

$$\{x_n\} \cdot \{y_n\} = \{x_n y_n\}$$

and

$$\{x_n\}^* = \{\overline{x_n}\}.$$

This makes c a commutative unital C^*-algebra, with unit $I = (1, 1, \ldots)$. Show that its maximal ideal space is

$$\mathcal{M}_c = \{\varphi_k : k \in \mathbb{N}\} \cup \{\varphi_\infty\}$$

where $\varphi_k(\{x_n\}) = x_k$ for $k \in \mathbb{N}$ and $\varphi_\infty(\{x_n\}) = \lim_{n \to \infty} x_n$.

Note that the map $k \to \varphi_k$ is a homeomorphism of $\mathbb{N}$ onto its range in $\mathcal{M}_c$, and clearly $\varphi_k \to \varphi_\infty$ weak* as $k \to \infty$. Thus $\mathcal{M}_c$ is naturally homeomorphic to the one-point compactification (see [33], p.183) of $\mathbb{N}$.

5.41. Consider ℓ^∞ as a commutative unital C^*-algebra with product $\{x_n\} \cdot \{y_n\} = \{x_n y_n\}$ and involution $\{x_n\}^* = \{\overline{x_n}\}$. The *cluster set at infinity* of $x = \{x_n\} \in \ell^\infty$ is the set

$$Cl_\infty(x) \equiv \{\lambda \in \mathbb{C} : \text{for every } \varepsilon > 0 \text{ and } N \in \mathbb{N},$$
$$\text{there exists } n \geq N \text{ with } |x_n - \lambda| < \varepsilon\}$$
$$= \{\lambda \in \mathbb{C} : \text{there exists a subsequence } \{x_{n_k}\} \text{ of }$$
$$\{x_n\} \text{ with } x_{n_k} \to \lambda \text{ as } k \to \infty\}.$$

Show that for any $x = \{x_n\} \in \ell^\infty$,

$$\sigma(x) = \{x_n : n \in \mathbb{N}\} \cup Cl_\infty(x) = \overline{\{x_n : n \in \mathbb{N}\}}.$$

5.42. Consider ℓ^∞ as a unital C^*-algebra as in the previous exercise. The *fiber at infinity* is the subset $X_\infty \subseteq \mathcal{M}_{\ell^\infty}$ defined by

$$X_\infty = \{\varphi \in \mathcal{M}_{\ell^\infty} : \varphi(x) = \lim_{n \to \infty} x_n \text{ for every } x = \{x_n\} \in c\}.$$

(a) For $n \in \mathbb{N}$, let $e_n = (0, \dots, 0, 1, 0 \dots)$ with a 1 in the nth position and 0's elsewhere. Show that

$$X_\infty = \{\varphi \in \mathcal{M}_{\ell^\infty} : \varphi(e_n) = 0 \text{ for all } n \in \mathbb{N}\}$$
$$= \{\varphi \in \mathcal{M}_{\ell^\infty} : c_0 \subseteq \ker \varphi\}.$$

(b) Suppose that for each $k \in \mathbb{N}$, $\varphi_k \in \mathcal{M}_{\ell^\infty}$ is defined by $\varphi_k(x) = x_k$ for $x = \{x_n\} \in \ell^\infty$. Show that

$$\mathcal{M}_{\ell^\infty} = \{\varphi_k : k \in \mathbb{N}\} \cup X_\infty.$$

(c) With the terminology of the last exercise, argue that for any $x \in \ell^\infty$,

$$Cl_\infty(x) = \{\hat{x}(\varphi) : \varphi \in X_\infty\}.$$

(d) Let $x \in \ell^\infty$. Show that $x \in c$ if and only if $\hat{x}$ is constant on X_∞.
(e) Show that $\{\varphi_k : k \in \mathbb{N}\}$ is dense in $\mathcal{M}_{\ell^\infty}$ in the weak* topology.
(f) Let $\varphi \in X_\infty$. Show that (the assertion of (e) notwithstanding) there is no subsequence $\{\varphi_{n_k}\}$ of $\{\varphi_n\}$ with $\varphi_{n_k} \to \varphi$ weak*.

5.43. Can a Banach limit on ℓ^∞ (see Exercise 3.9) be multiplicative?

5.44. Recall that in a unital C^*-algebra $\mathcal{A}$, the positive elements of $\mathcal{A}$ are defined to be those *self-adjoint* $A \in \mathcal{A}$ with $\sigma(A) \subseteq [0, \infty)$. Denote the collection of positive elements $\mathcal{A}_+$ and set $\mathcal{A}_- = \{A : -A \in \mathcal{A}_+\}$.

(a) Show that $\mathcal{A}_+ \cap \mathcal{A}_- = \{0\}$.
(b) Follow the outline below to show that every self-adjoint A in a C^*-algebra $\mathcal{A}$ can be written in the form $A = A_+ - A_-$, where A_+ and A_- are both positive elements of $\mathcal{A}$ and $A_+ A_- = A_- A_+ = 0$.
    Outline: Note that the identity function $h(t) = t$ on the real line can be written as $f - g$, where $f(t) = \max(0, t)$ and $g(t) = -\min(0, t)$. Observe that f, g are continuous nonnegative functions on the real line with $fg = 0$. For A self-adjoint, set $A_+ = f(A)$ and $A_- = g(A)$ as given by the functional calculus. Check that A_+ and A_- have the desired properties.

5.45. Suppose that $0 \le A \le B$ for self-adjoint elements A, B in a C^* algebra.

(a) Show that $B \le \|B\| I$. Hint: Consider $C^*(B) \cong C(\sigma(B))$ where B corresponds to the identity function on the spectrum of B. Use the functional calculus, with the function $f(x) = \|B\| - x$ on $\sigma(B)$.

(b) Show $\|A\| \leq \|B\|$. Hint: Consider $C^*(A) \cong C(\sigma(A))$ with A corresponding to the identity function on the spectrum of A. Use the functional calculus with the function $f(x) = \|B\| - x$.

(c) Show that $0 \leq A \leq B$ need not imply $A^2 \leq B^2$ by considering

$$X = \begin{bmatrix} 1 & 0 \\ 0 & 0 \end{bmatrix}$$

and

$$Y = \begin{bmatrix} \frac{1}{2} & \frac{1}{2} \\ \frac{1}{2} & \frac{1}{2} \end{bmatrix}.$$

Show $0 \leq X \leq X + Y$. Is $X^2 \leq (X + Y)^2$?

(d) Show that if $0 \leq A \leq B$ and A and B commute, then $A^n \leq B^n$ for every positive integer n. More generally, show that if there are positive elements C_j, $1 \leq j \leq k$, with

$$0 \leq A \leq C_1 \leq C_2 \leq \cdots \leq C_k \leq B$$

so that any two neighbors in this list commute, then $A^n \leq B^n$ for any positive integer n.

5.46. Suppose that P and Q are orthogonal projections onto closed subspaces M and N in $\mathcal{H}$, respectively. Show that $P \geq Q$ if and only if $N \subseteq M$.

5.47. Let $\mathcal{H}$ be a Hilbert space. An operator T in $\mathcal{B}(\mathcal{H})$ is said to be a *contraction* if $\|T\| \leq 1$.

(a) Show that T is a contraction if and only if $I - T^*T \geq 0$.

(b) Suppose that A and B are bounded linear operators on $\mathcal{H}$ with B invertible. Show that AB^{-1} is a contraction if and only if $A^*A \leq B^*B$.

5.48. What's wrong with the following "proof" that for an arbitrary element B of a unital C^*-algebra $\mathcal{A}$, the element B^*B is positive: Let $A = B^*B$. Clearly A is self-adjoint. Using the Gelfand transform we have

$$\Gamma(A) = \Gamma(B^*B) = \Gamma(B^*)\Gamma(B) = \overline{\Gamma(B)}\Gamma(B) = |\Gamma(B)|^2 \geq 0$$

so that $\sigma(A) = \sigma(\Gamma(A)) = \text{range } |\Gamma(B)|^2 \subseteq [0, \infty)$.

Chapter 6
The Spectral Theorem

> *Most students of mathematics learn quite early and most*
> *mathematicians remember till quite late that every Hermitian*
> *matrix (and in particular every real symmetric matrix) may be*
> *put into diagonal form.... The spectral theorem is widely and*
> *correctly regarded as the generalization of this assertion to*
> *operators on Hilbert space.*
> P. Halmos ([15], p. 241).

The literature of operator theory has a variety of dissimilar looking statements that get called "the spectral theorem." At their simplest, they describe either compact self-adjoint operators (as we did in Section 4.3) or compact normal operators on a Hilbert space. In either of these cases, a description of the operator connected to eigenvectors is still possible. In this chapter, we want to move to the more general case of bounded normal operators on a Hilbert space $\mathscr{H}$—operators that need not have any eigenvectors.

6.1 Normal Operators Are Multiplication Operators

We will begin with a formulation of the spectral theorem that says that bounded normal operators are "multiplication operators" when viewed in an appropriate way. This will give a statement of the spectral theorem which is particularly easy to digest and remember. Throughout this chapter we will work with *separable* Hilbert spaces.

To motivate what will become the statement of the spectral theorem it will be convenient to first look at the finite-dimensional situation. Here the usual formulation of the spectral theorem says that if A is a self-adjoint, or more generally a normal, $n \times n$ matrix with complex entries, then $\mathbb{C}^n$ has an orthonormal basis consisting of eigenvectors of A. If we denote such an orthonormal basis by $\{v_j\}_1^n$ and suppose that

$$Av_j = \alpha_j v_j,$$

then $(\alpha_1, \alpha_2, \ldots, \alpha_n)$ is an n-tuple of complex numbers. Think of this n-tuple as a function φ in $L^\infty(X, \mu)$, where $X \equiv \{1, 2, \ldots, n\}$, μ is counting measure on the subsets of X, and $\varphi(k) = \alpha_k$. Let $W : L^2(X, \mu) \to \mathscr{H} = \mathbb{C}^n$ be defined by

$$Wf = f(1)v_1 + f(2)v_2 + \cdots + f(n)v_n$$

and observe that W is a unitary map of $L^2(X, \mu)$ onto $\mathbb{C}^n$. Now suppose that M_φ is the operator of multiplication by φ acting on $L^2(X, \mu)$, and that f is in $L^2(X, \mu)$.

B.D. MacCluer, *Elementary Functional Analysis*, DOI 10.1007/978-0-387-85529-5_6, © Springer Science+Business Media, LLC 2009

Since $\varphi(k)$ is by definition the eigenvalue α_k for A, we have

$$
\begin{aligned}
WM_\varphi f &= W(\varphi f) \\
&= \varphi(1)f(1)v_1 + \cdots \varphi(n)f(n)v_n \\
&= f(1)\alpha_1 v_1 + \cdots + f(n)\alpha_n v_n \\
&= A(f(1)v_1 + \cdots + f(n)v_n) \\
&= AW f.
\end{aligned}
$$

It follows that $M_\varphi = W^{-1}AW$. This argument can be generalized to any normal $T \in \mathcal{B}(\mathcal{H})$, *provided* T has enough eigenvectors to form an orthonormal basis for $\mathcal{H}$; see Exercise 6.1. The problem, of course, is that T need not have *any* eigenvectors. Nevertheless, this point of view suggests we focus on the following concept.

Definition 6.1. A bounded linear operator A on a separable Hilbert space $\mathcal{H}$ is *unitarily equivalent to a multiplication* if there is a σ-finite measure space (X,μ), a function $\varphi \in L^\infty(X,\mu)$, and a unitary $W : L^2(X,\mu) \to \mathcal{H}$ such that

$$
WM_\varphi = AW,
$$

where M_φ denotes the operator of multiplication by φ on the (necessarily separable) Hilbert space $L^2(X,\mu)$.

In the finite-dimensional example just described, X is $\{1,2,\ldots,n\}$ and μ is counting measure, so that the norms on $L^2(X,\mu)$ and $L^\infty(X,\mu)$ are given by

$$
\|f\|_2 = \left(\sum_{j=1}^{n} |f(j)|^2 \right)^{1/2}
$$

and

$$
\|\varphi\|_\infty = \max_{1 \le j \le n} |\varphi(j)|,
$$

respectively. The symbol of the multiplication operator here is the $L^\infty(X)$ function φ.

Notice that only normal operators can be unitarily equivalent to a multiplication, since all multiplication operators are normal. The spectral theorem says the converse:

Theorem 6.2 (Spectral Theorem, Multiplication Version). *Every normal operator on a separable Hilbert space is unitarily equivalent to a multiplication operator.*

The goal of this section is to prove Theorem 6.2. Our presentation is influenced by that in [2] and in [15], from which the quote that introduces this chapter is taken. We will first prove Theorem 6.2 under an additional hypothesis, involving the notion of cyclicity, a concept that was briefly explored in Exercise 2.22 of Chapter 2.

Definition 6.3. A vector h in $\mathcal{H}$ is a *cyclic* vector for the operator A in $\mathcal{B}(\mathcal{H})$ if (finite) linear combinations of the vectors in the orbit of h, namely

$$\{h, Ah, A^2h, A^3h, \dots\},$$

are dense in $\mathscr{H}$. Equivalently, h is a cyclic vector for A if

$$\{p(A)h : p \text{ is a polynomial}\}$$

is dense in $\mathscr{H}$.

As an example, note that the forward shift S on $\ell^2(\mathbb{N})$ has cyclic vector $h = (1,0,0,\dots)$ (among others). Not every operator has a cyclic vector (the identity operator on a space of dimension greater than one is an easy example). The next example, which is generalized in Exercise 6.2, shows how, for a diagonal matrix, an eigenvalue of multiplicity greater than 1 prohibits cyclicity.

Example 6.4. In this example we look at the linear operator on $\mathbb{C}^3$ given by the matrix

$$A = \begin{bmatrix} 1 & 0 & 0 \\ 0 & 1 & 0 \\ 0 & 0 & 2 \end{bmatrix}.$$

Note that for any polynomial p and any vector $h \in \mathbb{C}^3$, the first two components of $p(A)h$ agree. Is there a choice of h so that h is cyclic for A? Fix a column vector $h = (h_1, h_2, h_3)^t$. First observe that no h_j, $j = 1, 2, 3$ could be zero if h is a cyclic vector, since the corresponding component of $p(A)h$ would be zero for all polynomials p. Next, the necessarily nonzero vector $(\overline{h_2}, -\overline{h_1}, 0)^t$ is orthogonal to $p(A)h$ for all p. Thus

$$\{p(A)h : p \text{ a polynomial}\}$$

is not dense in $\mathbb{C}^3$, and we conclude that A has no cyclic vector.

One can rephrase the invariant subspace problem in terms of cyclic vectors: Since the closure of the span of $\{h, Ah, A^2h, \dots\}$ is clearly an A-invariant subspace, and the smallest closed A-invariant subspace containing a given vector h must contain the closure of the span of its orbit under A, the question becomes whether every bounded linear operator has a nonzero, noncyclic vector. Enflo's construction of a Banach space operator with no invariant subspace at its heart proceeds by constructing a Banach space on which a simple multiplication operator has no nonzero, noncyclic vector.

Recall that when A is a normal operator in $\mathscr{B}(\mathscr{H})$, the C^*-algebra $C^*(A)$ is commutative, and that it is the closure of the polynomials in A and A^*. We will say that $C^*(A)$ has a cyclic vector h if $\{Bh : B \in C^*(A)\}$ is dense in $\mathscr{H}$; that is, if

$$\{p(A, A^*)h : p = p(z, w) \text{ is a two-variable polynomial}\}$$

is dense in $\mathscr{H}$. Note that if A has a cyclic vector, then trivially so does $C^*(A)$, and for self-adjoint A, A has cyclic vector h if and only if $C^*(A)$ has cyclic vector h.

Most of the work in proving the spectral theorem will be done in verifying a reduced form of the result, which we state next.

Theorem 6.5. *Let A be a normal operator in $\mathscr{B}(\mathscr{H})$, where $\mathscr{H}$ is a separable Hilbert space. If the C^*-algebra $C^*(A)$ has a cyclic vector, then the operator A is unitarily equivalent to a multiplication.*

Before proceeding to the proof of Theorem 6.5 we make a few comments about what is to be done. We will need to construct a σ-finite measure space (X, μ) and a unitary operator $W : L^2(X, \mu) \to \mathscr{H}$ such that $W^{-1}AW$ is a multiplication operator on $L^2(X, \mu)$. Are there any likely candidates for this measure space, given that our "starting data" is the normal operator A? Based on our previous experience (for example, Theorem 5.46 and the discussion preceding it), $\sigma(A)$ might be a reasonable candidate for X. It's less clear where the measure μ is to come from; we will see that the Riesz–Markov theorem, as described in Exercise 3.7 in Chapter 3 and Section A.5 of the Appendix, will produce the measure μ. Moreover, we will see that the multiplication operator we construct in the proof of Theorem 6.5 can be taken to be M_z, the operator of multiplication by the identity function, acting on $L^2(X, \mu)$ for $X = \sigma(A)$.

The first line of the proof of Theorem 6.5 will say "fix a cyclic vector h in $\mathscr{H}$." The fact that h is cyclic for $C^*(A)$ only plays a role at the very end of the proof, so the initial steps in the proof hold for any fixed vector h in $\mathscr{H}$. The reader may find it helpful to delay including the cyclicity hypothesis until the end, where its role becomes transparent.

Proof (Theorem 6.5). Fix a cyclic vector h in $\mathscr{H}$. Let $X = \sigma(A)$, the spectrum of A, which is a compact subset of $\mathbb{C}$. We know by Theorem 5.46 that there is a unique isometric $*$-isomorphism γ between $C^*(A)$ and $C(\sigma(A))$ sending A to z, the identity function on $\sigma(A)$. Properties of this $*$-isomorphism are indicated below; here p is a polynomial in two variables and f, g denote arbitrary functions in $C(\sigma(A))$.

$$C^*(A) \underset{\gamma^{-1}}{\overset{\gamma}{\rightleftarrows}} C(\sigma(A))$$

$$I \longleftrightarrow 1$$

$$A \longleftrightarrow z$$

$$A^* \longleftrightarrow \bar{z}$$

$$p(A, A^*) \longleftrightarrow p(z, \bar{z})$$

$$f(A) \longleftrightarrow f$$

$$f(A)^* \longleftrightarrow \bar{f}$$

$$f(A)g(A) \longleftrightarrow fg$$

$$f(A) + g(A) \longleftrightarrow f + g$$

Define Λ on $C(X) = C(\sigma(A))$ by

$$\Lambda(f) = \langle f(A)h, h \rangle,$$

where $f(A)$ is defined, as above, by the continuous functional calculus and h is our chosen fixed vector. We claim that Λ is a bounded, positive linear functional on $C(\sigma(A))$. Linearity follows from the fact that $f(A) + g(A) = (f+g)(A)$ and $(\alpha f)(A) = \alpha f(A)$ for any scalar α. Since γ is an isometry,

$$|\Lambda(f)| = |\langle f(A)h, h\rangle| \leq \|f(A)h\| \, \|h\| \leq \|f(A)\| \, \|h\|^2 = \|f\|_\infty \|h\|^2,$$

so Λ is bounded, with norm at most $\|h\|^2$.

Positivity of Λ is verified as follows. A nonnegative function f in $C(\sigma(A))$ can be written as $f = g^2$ for some nonnegative g in $C(\sigma(A))$. For such f,

$$\Lambda(f) = \Lambda(g^2) = \langle g^2(A)h, h\rangle = \langle g(A)g(A)h, h\rangle = \langle g(A)h, g(A)h\rangle = \|g(A)h\|^2 \geq 0,$$

where we have used the fact that $g(A)$ is self-adjoint in $\mathscr{B}(\mathscr{H})$, since g is self-adjoint (real-valued) in $C(\sigma(A))$.

Now we invoke the Riesz–Markov theorem for $C(X)$ (see Section A.5 of the Appendix) and conclude that the positive linear functional Λ is given by integration against a unique positive finite Borel measure μ on $X = \sigma(A)$:

$$\langle f(A)h, h\rangle = \int_{\sigma(A)} f d\mu$$

for all $f \in C(\sigma(A))$. The μ-measure of $\sigma(A)$ is equal to the norm of the linear functional Λ.

Next we want a unitary map $W : L^2(X, \mu) \to \mathscr{H}$ so that $W^{-1}AW$ is a multiplication operator. We will first define W on $C(\sigma(A))$, instead of all of $L^2(\sigma(A), \mu)$. For f continuous on $\sigma(A)$, set $Wf = f(A)h$. It is easy to check that W is linear, using the linearity of the $*$-isomorphism γ. Moreover, for f and g continuous on $\sigma(A)$,

$$\begin{aligned}
\langle Wf, Wg\rangle_{\mathscr{H}} &= \langle f(A)h, g(A)h\rangle_{\mathscr{H}} \\
&= \langle g(A)^* f(A)h, h\rangle_{\mathscr{H}} \\
&= \langle (\bar{g}f)(A)h, h\rangle_{\mathscr{H}} \\
&= \Lambda(\bar{g}f) \\
&= \int_{\sigma(A)} \bar{g}f d\mu \\
&= \langle f, g\rangle_{L^2(X, \mu)}.
\end{aligned}$$

This calculation shows that W is a linear isometry of $C(X)$, *equipped with the $L^2(X, \mu)$ norm*, into $\mathscr{H}$. Since $C(X)$ is dense in $L^2(X, \mu)$, we can extend W from $C(X)$ to the Hilbert space $L^2(X, \mu)$ by continuity in a unique way. It is easy to check that this extension, which we continue to denote by W, is still a linear isometry into $\mathscr{H}$. What is the range of W? Since W is an isometry, it has closed range, and thus the range must be a closed subspace of $\mathscr{H}$ which contains

$$\{f(A)h : f \in C(X)\} = \{Bh : B \in C^*(A)\}.$$

Here the cyclicity hypothesis enters. By assumption, $\{Bh : B \in C^*(A)\}$ is dense in $\mathscr{H}$, and hence W is a unitary map of $L^2(X, \mu)$ *onto* $\mathscr{H}$.

Finally we verify that $W^{-1}AW$ is a multiplication operator on $L^2(X, \mu)$. We will actually show that

$$W^{-1}f(A)W = M_f \tag{6.1}$$

for all f in $C(X)$. In particular, this shows that $W^{-1}AW = M_z$, where z denotes the identity function on $X = \sigma(A)$. To verify Equation (6.1), compute, for $g \in C(X)$,

$$WM_f g = W(fg) = (fg)(A)h = f(A)g(A)h = f(A)Wg,$$

so that $WM_f = f(A)W$ on the dense subset $C(X)$ in $L^2(X, \mu)$, and thus also on all of $L^2(X, \mu)$. This completes the proof. $\qquad\qquad\qquad\qquad\qquad\qquad\square$

We next want to see how to remove the cyclicity hypothesis from Theorem 6.5. Very roughly, the idea is that if $C^*(A)$ doesn't have a cyclic vector, we may write $\mathscr{H}$ as a direct sum of subspaces $\mathscr{H}_n$, each of which is invariant under $\mathscr{A} = C^*(A)$ (meaning $B\mathscr{H}_n \subseteq \mathscr{H}_n$ for all $B \in C^*(A)$) and such that the restriction of A to each piece does have the desired cyclicity property. We then apply the reduced form of the spectral theorem to the pieces and use the resulting measure spaces and unitary maps to build a unitary equivalence of A to a multiplication operator.

To carry out the details, we begin by reviewing the notion of the direct sum of Hilbert spaces, and the direct sum of operators (see Exercises 1.31 in Chapter 1 and 2.13 in Chapter 2). Given a finite or countable collection of pairwise orthogonal subspaces $\mathscr{H}_n$ of a Hilbert space, we denote the set of all convergent sums $\sum h_n$ with $h_n \in \mathscr{H}_n$ by $\sum \oplus \mathscr{H}_n$. This is isomorphic to the set of all sequences (or k-tuples in the finite case) $(h_1, h_2, \ldots)$ where $h_n \in \mathscr{H}_n$ and $\sum \|h_n\|^2 < \infty$ under the map

$$(h_1, h_2, \ldots) \to h_1 + h_2 + \cdots.$$

When convenient we will think of $\sum \oplus \mathscr{H}_n$ as this set of sequences (or k-tuples) with inner product

$$\langle h, g \rangle = \sum \langle h_n, g_n \rangle.$$

Given operators $A_n \in \mathscr{B}(\mathscr{H}_n)$ with

$$\sup_n \|A_n\| < \infty$$

we define A on $\sum \oplus \mathscr{H}_n$ by

$$A(h_1, h_2, \ldots) = (A_1 h_1, A_2 h_2, \ldots).$$

It is easy to check that A is bounded on $\sum \oplus \mathscr{H}_n$, and that $\|A\|$ coincides with the supremum above. We call A the *direct sum* of the operators A_n, with respect to the decomposition $\sum \oplus \mathscr{H}_n$, and write

$$A = A_1 \oplus A_2 \oplus \cdots = \sum_n \oplus A_n.$$

Lemma 6.6. *Let A be a normal operator on a separable Hilbert space $\mathscr{H}$. There is a finite or countable collection of nonzero, pairwise orthogonal subspaces $\mathscr{H}_n$ of $\mathscr{H}$ satisfying*

(a) $\quad \mathscr{H} = \mathscr{H}_1 \oplus \mathscr{H}_2 \oplus \cdots$.

(b) $\quad$ *Each $\mathscr{H}_n$ is invariant under $\mathscr{A} = C^*(A)$, meaning $B\mathscr{H}_n \subseteq \mathscr{H}_n$ for each $B \in C^*(A)$; equivalently, each $\mathscr{H}_n$ is a reducing subspace for A.*

(c) $\quad$ *Each $\mathscr{H}_n$ contains a vector h_n which is cyclic for $\mathscr{A}_n \equiv C^*(A|_{\mathscr{H}_n})$; that is, $\{Bh_n : B \in C^*(A|_{\mathscr{H}_n})\}$ is dense in $\mathscr{H}_n$.*

Proof. We use a Zorn's lemma argument. Consider the following family $\mathscr{F}$. An element of $\mathscr{F}$ is a collection of nonzero pairwise orthogonal, $\mathscr{A}$-invariant closed subspaces $\mathscr{H}_\alpha$ of $\mathscr{H}$, each containing a cyclic vector for $C^*(A|_{\mathscr{H}_\alpha})$. The collection $\mathscr{F}$ is nonempty since if we pick any nonzero vector h in $\mathscr{H}$, the closure of $\{\mathscr{A}h\}$ is an $\mathscr{A}$-invariant closed subspace of $\mathscr{H}$ with a cyclic vector. Partially order $\mathscr{F}$ by inclusion. Every totally ordered chain τ in $\mathscr{F}$ has an upper bound in $\mathscr{F}$, namely the union of all elements in τ. By Zorn's lemma there is a maximal element

$$\{\mathscr{H}_\alpha : \alpha \in I\}$$

in $\mathscr{F}$. Since $\mathscr{H}$ is separable and the $\mathscr{H}_\alpha$ are pairwise orthogonal, the index set I is finite or countable, and we write our maximal element as

$$\{\mathscr{H}_1, \mathscr{H}_2, \ldots\}.$$

We are done if we can show that

$$\mathscr{H} = \mathscr{H}_1 \oplus \mathscr{H}_2 \oplus \cdots.$$

If not, the direct sum on the right is a proper closed subspace $\mathscr{K}$ of $\mathscr{H}$, which is easily seen to be invariant under A and A^*. Its orthogonal complement, $\mathscr{K}^\perp$ is nonzero and orthogonal to each $\mathscr{H}_1, \mathscr{H}_2, \ldots$. Pick a nonzero vector in $\mathscr{K}^\perp$, say ζ, and look at the cyclic subspace it generates,

$$\mathscr{H}_0 \equiv \{\mathscr{A}\zeta\}^- \subseteq \mathscr{K}^\perp.$$

Now $\{\mathscr{H}_0, \mathscr{H}_1, \mathscr{H}_2, \ldots\}$ is in $\mathscr{F}$, contradicting the maximality of $\{\mathscr{H}_1, \mathscr{H}_2, \ldots\}$. $\quad\square$

The next lemma shows that a direct sum of a sequence (or finite collection) of operators, each of which is unitarily equivalent to a multiplication, is itself unitarily equivalent to a multiplication. It uses the notion of the *disjoint union* of an at most countable collection of sets. This is defined as follows: Given sets X_n let

$$\tilde{X}_n = X_n \times \{n\},$$

so that $\tilde{X}_j \cap \tilde{X}_k = \emptyset$ if $j \neq k$, and let

$$X = \cup_n \tilde{X}_n.$$

The parameter n in an element $(x, n) \in \tilde{X}_n$ is there merely to make the sets $\tilde{X}_n$ and $\tilde{X}_m$ disjoint, even when X_n and X_m have points in common. If, as will be our case, we have measures μ_n so that (X_n, μ_n) is a measure space for each n, we can define a copy $\tilde{\mu}_n$ of μ_n on $\tilde{X}_n$ by

$$\tilde{\mu}_n(E \times \{n\}) \equiv \mu_n(E)$$

for each μ_n-measurable subset E of X_n. A measurable set in X is a set F for which $F \cap \tilde{X}_n$ is $\tilde{\mu}_n$-measurable for each n; such sets form a σ-algebra on X. We define a measure μ on X by

$$\mu(F) = \tilde{\mu}_1(F \cap \tilde{X}_1) + \tilde{\mu}_2(F \cap \tilde{X}_2) + \cdots \tag{6.2}$$

for each measurable $F \subseteq X$. If μ_n is σ-finite for each n, then μ will be σ-finite also.

Lemma 6.7. *If $A_1, A_2, \ldots$ is a finite or countable collection of operators on, respectively, separable Hilbert spaces $\mathcal{H}_1, \mathcal{H}_2, \ldots$, with each A_n unitarily equivalent to a multiplication and $\sup_n \|A_n\| < \infty$, then $\sum \oplus A_n$, acting on $\sum \oplus \mathcal{H}_n$, is unitarily equivalent to a multiplication.*

Proof. We are given the existence of σ-finite measure spaces (X_n, μ_n), unitary operators $W_n : L^2(X_n, \mu_n) \to \mathcal{H}_n$, and functions f_n in $L^\infty(X_n, \mu_n)$ such that

$$W_n M_{f_n} = A_n W_n$$

for each n. Moreover, $\|A_n\| = \|M_{f_n}\| = \|f_n\|_\infty$, so that $\sup_n \|f_n\|_\infty$ is finite. Let X be the disjoint union of the sets X_n, as described above, and define μ on X by Equation (6.2) for all F with the property that $F \cap \tilde{X}_n$ is $\tilde{\mu}_n$-measurable for each n. It is easy to verify that $L^2(X, \mu)$ is isometrically isomorphic to $\sum \oplus L^2(X_n, \mu_n)$ via the map

$$g \in L^2(X, \mu) \to (g_1, g_2, \ldots) \text{ where } g_n \equiv g|_{\tilde{X}_n}.$$

Set $W = \sum \oplus W_n$, where W_n is our unitary map from $L^2(X_n, \mu_n)$ onto $\mathcal{H}_n$. We leave it to the reader to check that W is a linear surjection from $\sum \oplus L^2(X_n, \mu_n) \cong L^2(X, \mu)$ onto $\sum \oplus \mathcal{H}_n$ which is isometric, so that W is unitary.

Finally, define the $\mathbb{C}$-valued function f on X by $f(x, n) = f_n(x)$ for $x \in X_n$. Observe that f is in $L^\infty(X, \mu)$ since $\sup_n \|f_n\|_{L^\infty(X_n, \mu_n)} < \infty$. Thus M_f is a bounded linear operator on $L^2(X, \mu)$ and we claim that

$$WM_f = \left(\sum \oplus A_j \right) W,$$

which is the statement that $\sum \oplus A_j$ acting on $\sum \oplus \mathcal{H}_j$ is unitarily equivalent to a multiplication. To see this, let $g \in L^2(X, \mu)$ with $g = (g_1, g_2, \ldots)$ in $\sum \oplus L^2(X_n, \mu_n)$ and compute

$$(WM_f)g = W(fg) = W(f_1 g_1, f_2 g_2, \ldots)$$
$$= (W_1(f_1 g_1), W_2(f_2 g_2), \ldots)$$

$$= (A_1 W_1 g_1, A_2 W_2 g_2, \ldots)$$
$$= \left(\sum \oplus A_j \right) W g,$$

as desired. □

We are now ready to prove Theorem 6.2, the full version of the spectral theorem, in its "multiplication form."

Proof (Theorem 6.2). Let $A \in \mathcal{B}(\mathcal{H})$ be normal and set $\mathscr{A} = C^*(A)$. By Lemma 6.6 we may write $\mathcal{H} = \sum \oplus \mathcal{H}_n$ for pairwise orthogonal nonzero $\mathcal{H}_n$, where $\mathscr{A} \mathcal{H}_n \subseteq \mathcal{H}_n$ and each $\mathcal{H}_n$ has an $\mathscr{A}_n$-cyclic vector. By Theorem 6.5, the restriction A_n of A to $\mathcal{H}_n$ is unitarily equivalent to a multiplication. Since $\|A\| < \infty$, $\sup_n \|A_n\| < \infty$. Thus by Lemma 6.7, $\sum \oplus A_n$ acting on $\sum \oplus \mathcal{H}_n$ is unitarily equivalent to a multiplication, so that A acting on $\mathcal{H}$ is also (where we identify $\mathcal{H}$ and $\sum \oplus \mathcal{H}_n$). □

Since the measure μ constructed in the proof of Theorem 6.5 is finite, the underlying measure space in Theorem 6.2 is at least σ-finite. In Exercise 6.3 you are asked to show that in fact this measure space can be taken to be finite. We have used the separability assumption on $\mathcal{H}$ here; without it we would not know in Lemma 6.6 that the collection of subspaces is at most countable. There is a version of the spectral theorem for nonseparable Hilbert spaces, which follows along similar lines, but requires a bit more care.

6.2 Spectral Measures

In this section we explore the "spectral measure" version of the spectral theorem. This is its more classical articulation, but it will require some effort to develop the somewhat peculiar notion of spectral measures.

When A is a normal operator on $\mathcal{B}(\mathcal{H})$, for $\mathcal{H}$ a separable Hilbert space, we have used the continuous functional calculus to define $f(A)$, for any continuous function f on $\sigma(A)$, as schematically illustrated by

$$\mathcal{B}(\mathcal{H}) \supseteq C^*(A) \underset{\gamma^{-1}}{\overset{\gamma}{\rightleftarrows}} C(\sigma(A))$$
$$A \longleftrightarrow z$$
$$f(A) \longleftrightarrow f$$

Next we will show how to use the spectral theorem to extend the continuous functional calculus to define $g(A)$ to be an operator in $\mathcal{B}(\mathcal{H})$, when g is any bounded Borel measurable function on $\sigma(A)$. By Theorem 6.2, A is unitarily equivalent to M_φ, the operator of multiplication by φ, on $L^2(Y, \mu)$ for some σ-finite measure space Y and some $\varphi \in L^\infty(Y, \mu)$. Thus there is a unitary operator $W : L^2(Y, \mu) \to \mathcal{H}$ such that $A = W M_\varphi W^{-1}$. As a provisional definition, we propose setting

$$g(A) = W M_{g \circ \varphi} W^{-1}. \tag{6.3}$$

(Later we will give another definition, which is independent of the unitary equivalence of A to a multiplication, and see that it agrees with our provisional definition above.) To see that our provisional definition makes sense, we have one issue to address. Can we choose a representative for φ whose range is contained in $\sigma(A)$? If so, then $g \circ \varphi$ will be defined. We know that

$$\text{ess range } \varphi = \sigma(M_\varphi) = \sigma(A)$$

where "ess range φ" denotes the essential range of φ (see Exercise 5.24 in Chapter 5). So our question is whether, altering φ on a set of μ-measure zero if necessary, we have range $\varphi \subseteq \sigma(A)$. If λ is not in the essential range of φ, there exists $\varepsilon > 0$ so that the inverse image under φ of the disk centered at λ with radius ε has μ-measure zero. The complement of $\sigma(A)$ is an open set in $\mathbb{C}$, and any open cover of $\mathbb{C} \backslash \sigma(A)$ has a countable subcover ([33], p. 191). Thus $\mathbb{C} \backslash \sigma(A)$ may be covered by a countable collection of open disks whose inverse images under φ have μ-measure zero, and therefore $\varphi^{-1}(\mathbb{C} \backslash \sigma(A))$ has μ-measure zero. Thus we may alter φ on a set of measure zero so that its range is contained in $\sigma(A)$, as desired, and our proposed definition of "$g(A)$" in Equation (6.3) is justified.

This definition of $g(A)$ for g a bounded Borel measurable function on $\sigma(A)$ is called the "Borel functional calculus," and we claim that it extends the continuous functional calculus. To verify this, note that when $g(z) = z$, then Equation (6.3) gives $g(A) = A$, and when $g(z) = \bar{z}$, Equation (6.3) identifies $g(A)$ as $W M_{\bar{\varphi}} W^{-1} = A^*$. Now apply the uniqueness statement in Theorem 5.46. We leave the details to the reader.

A particular instance of the Borel functional calculus comes from choosing $g = \chi_S$, the characteristic function of a Borel subset S of $\sigma(A)$. Observe that $(\chi_S \circ \varphi)(x) = 1$ if $\varphi(x)$ is in S, and is equal to 0 otherwise. Thus

$$\chi_S \circ \varphi = \chi_{\varphi^{-1}(S)}$$

and we have

$$\chi_S(A) = W M_{\chi_{\varphi^{-1}(S)}} W^{-1}.$$

What kind of operator is this? Since any characteristic function χ is an idempotent (meaning $\chi^2 = \chi$), we have

$$[\chi_S(A)]^2 = \chi_S(A).$$

Clearly, $\chi_S(A)$ is self-adjoint. Recall (Exercise 2.17 in Chapter 2) that any operator T in $\mathscr{B}(\mathscr{H})$ that is self-adjoint and satisfies $T^2 = T$ is an orthogonal projection (onto its range). Thus the mapping that sends each Borel subset S of $\sigma(A)$ to $\chi_S(A)$ associates to each such S an orthogonal projection in $\mathscr{B}(\mathscr{H})$. The identity $I \in \mathscr{B}(\mathscr{H})$ is associated to the set $\sigma(A)$, and the zero operator is associated to the empty set. It is easy to see that orthogonal projections onto mutually orthogonal subspaces of $\mathscr{H}$ are associated to disjoint sets. To continue this line of investigation further, we make the following definition. Recall that by a measurable space $(X, \mathscr{F})$ we mean a set X together with a specific σ-algebra $\mathscr{F}$ of subsets of X.

Definition 6.8. Let $(X, \mathscr{F})$ be any measurable space. A *spectral measure* on X is a function $E : \mathscr{F} \to \mathscr{B}(\mathscr{H})$, where $\mathscr{H}$ is a Hilbert space, satisfying the following:

(1) For each S in $\mathscr{F}$, $E(S)$ is an orthogonal projection.
(2) $E(\emptyset) = 0$ and $E(X) = I$.
(3) If S_1, S_2 are in $\mathscr{F}$, and $S_1 \cap S_2 = \emptyset$, then

$$E(S_1)\mathscr{H} \perp E(S_2)\mathscr{H}.$$

(4) If $\{S_k\}_1^\infty$ is a sequence of pairwise disjoint sets from $\mathscr{F}$, then for each $h \in \mathscr{H}$

$$\sum_{k=1}^n E(S_k)h \to E(\cup_{k=1}^\infty S_k)h$$

as $n \to \infty$.

Note that as a special case of (4), if $S_1 \cap S_2 = \emptyset$, then $E(S_1) + E(S_2) = E(S_1 \cup S_2)$.

Typically our interest in spectral measures will be in the case that X is $\mathbb{C}$ or a subset of $\mathbb{C}$, and $\mathscr{F}$ is the Borel σ-algebra on X. There are some useful "ordinary" measures, defined on the σ-algebra $\mathscr{F}$, which can be created from a spectral measure E; we will see this in Proposition 6.14 below.

Example 6.9. To find an easy example of a spectral measure, let $(X, \mathscr{F}, \mu)$ be a measure space and set $\mathscr{H} = L^2(X, \mu)$. Define $E : \mathscr{F} \to \mathscr{B}(\mathscr{H})$ by $E(S) = M_{\chi_S}$, the operator of multiplication by the characteristic function χ_S, acting on $L^2(X, \mu)$; the function E is a spectral measure on X. The reader is encouraged to check the details verifying properties (1)–(4) in Definition 6.8.

Example 6.10. Suppose that T is a diagonal operator on ℓ^2 with diagonal sequence $\{\lambda_1, \lambda_2, \dots\}$. Define E on the Borel subsets of $\mathbb{C}$ by setting $E(S)$ to be the diagonal operator with diagonal sequence $\{\alpha_1, \alpha_2, \dots\}$, where $\alpha_j = 1$ if λ_j is in S and $\alpha_j = 0$ if λ_j is not in S. One easily checks that E is a spectral measure.

Example 6.11. Suppose that M_φ is a multiplication operator on $L^2(X, \mu)$, where (X, μ) is a measure space. The function $E : \mathscr{F} \to \mathscr{B}(L^2(X, \mu))$ given by

$$E(S) = M_{\chi_{\varphi^{-1}(S)}}$$

for S a Borel subset of $\mathbb{C}$ is a spectral measure on $\mathbb{C}$. Clearly conditions (1) and (2) of Definition 6.8 hold. For (3) observe that if $S_1 \cap S_2 = \emptyset$, then for any $h_1, h_2 \in L^2(X, \mu)$ we have

$$\langle E(S_1)h_1, E(S_2)h_2 \rangle = \langle M_{\chi_{\varphi^{-1}(S_1)}}h_1, M_{\chi_{\varphi^{-1}(S_2)}}h_2 \rangle. \tag{6.4}$$

Disjointness of S_1 and S_2 ensures that $\chi_{\varphi^{-1}(S_1)}\chi_{\varphi^{-1}(S_2)} = 0$, so that the inner product in Equation (6.4) is zero. Property (4) in Definition 6.8 is the statement that for each $h \in L^2(X, \mu)$,

$$\chi_{\varphi^{-1}(\cup_1^n S_k)} h \to \chi_{\varphi^{-1}(\cup_1^\infty S_k)} h$$

in $L^2(X, \mu)$ as $n \to \infty$. This is easily verified by, say, an appeal to the dominated convergence theorem.

Notice that if S is contained in the complement of the range of φ, then $E(S) = 0$ by definition. Sometimes one defines the support of a spectral measure on $\mathbb{C}$ to be the complement of the union of all open sets S for which $E(S) = 0$. In this example, the support of the spectral measure E is the essential range of φ, or equivalently the spectrum of the operator M_φ (Exercise 6.5). We may think of the spectral measure E in this example as being defined on the Borel subsets of $\sigma(M_\varphi)$, and extended to all Borel subsets of $\mathbb{C}$ by setting $E(C) = 0$ if $C \subseteq \mathbb{C} \backslash \sigma(M_\varphi)$.

Example 6.12. Suppose $(X, \mathscr{F})$ is a measurable space and $E : \mathscr{F} \to \mathscr{B}(\mathscr{H})$ is a spectral measure. If $\mathscr{K}$ is another Hilbert space and $W : \mathscr{H} \to \mathscr{K}$ is unitary, then the formula

$$F(S) = WE(S)W^{-1}$$

defines a spectral measure $F : \mathscr{F} \to \mathscr{B}(\mathscr{K})$. The reader is asked to verify the details in Exercise 6.6.

Property (4) in Definition 6.8 is sometimes described by saying that

$$\sum_{k=1}^\infty E(S_k) = E(\cup_{k=1}^\infty S_k)$$

with the convergence of the sum of projections taking place in the *strong operator topology*. This terminology appeared earlier, in Exercise 2.23 of Chapter 2. Recall from that exercise that a sequence $\{T_n\}$ of operators on a Hilbert space is said to converge in the strong operator topology to an operator T if for each vector h, $T_n h \to Th$. The next result serves to describe the operator $E(\cup_{k=1}^\infty S_k) = \sum_1^\infty E(S_k)$. Recall that $\vee M_k$ denotes the closed linear span of the sets M_k, and when the M_k are pairwise orthogonal closed subspaces, $\vee M_k = \sum \oplus M_k$ (Exercise 1.33 in Chapter 1).

Lemma 6.13. *Suppose that $\mathscr{H}$ is a Hilbert space and that $\{E_k\}_{k=1}^\infty$ is a sequence of orthogonal projections on $\mathscr{H}$ with $E_j \mathscr{H} \perp E_k \mathscr{H}$ for all $j \neq k$. We have*

$$\sum_{k=1}^\infty E_k = E,$$

in the sense of strong operator convergence, where E is the projection of $\mathscr{H}$ onto

$$\bigvee_1^\infty E_k \mathscr{H} = \sum_1^\infty \oplus E_k \mathscr{H}.$$

Proof. Let $P_n = \sum_1^n E_k$. We want to show that $P_n h \to Eh$ for each $h \in \mathscr{H}$, where E is defined as in the statement of the lemma. First suppose that h is in the subspace $\sum_1^\infty \oplus E_k \mathscr{H}$, so that $h = \sum_1^\infty h_k$ with $h_k \in E_k \mathscr{H}$ and $\sum \|h_k\|^2 < \infty$. Since $P_n h = \sum_1^n h_k$, we have $P_n h \to h = Eh$ as $n \to \infty$. If, on the other hand, $h \perp E_k \mathcal{H}$ for all k, then $P_n h = 0 = Eh$. Given any $x \in \mathcal{H}$, write $x = y + z$ where $y \in \sum_1^\infty \oplus E_k \mathcal{H}$ and $z \perp E_k \mathcal{H}$ for all k. We have $P_n x \to y = Ex$, as desired. $\qquad\square$

By this lemma, we see that in (4) of Definition 6.8, $\sum_{k=1}^\infty E(S_k)$ is the projection onto $\sum_1^\infty \oplus E(S_k) \mathcal{H}$, and moreover, for each $h \in \mathcal{H}$,

$$\| E(\cup_1^\infty S_k) h \|^2 = \sum_1^\infty \| E(S_k) h \|^2. \tag{6.5}$$

Since a spectral measure E maps disjoint sets to projections with orthogonal ranges, it follows easily that

$$S_1 \subseteq S_2 \implies E(S_1) \mathcal{H} \subseteq E(S_2) \mathcal{H}.$$

To see this, write S_2 as the disjoint union of S_1 and $S_2 \setminus S_1$, so that

$$E(S_2) \mathcal{H} = E(S_1) \mathcal{H} \oplus E(S_2 \setminus S_1) \mathcal{H} \supseteq E(S_1) \mathcal{H}.$$

Using this observation we see that for any two measurable sets B_1 and B_2,

$$E(B_1 \cap B_2) = E(B_1) E(B_2),$$

and so the operators $E(B_1)$ and $E(B_2)$ commute. The details of this statement are left to the reader in Exercise 6.7. Since we have $E(S_k) \mathcal{H} \subseteq E(\cup_1^\infty S_j) \mathcal{H}$ for each k, and thus

$$\sum_1^\infty \oplus E(S_k) \mathcal{H} \subseteq E(\cup_1^\infty S_k) \mathcal{H},$$

the significance of condition (4) in Definition 6.8 is to demand the reverse containment.

The property in condition (4) of Definition 6.8 is certainly reminiscent of the defining property for a (scalar-valued) measure (i.e., countable additivity), and our next result describes constructing ordinary (complex) measures on $(X, \mathcal{F})$ from spectral measures. Here we return to our general setting of an arbitrary measurable space $(X, \mathcal{F})$.

Proposition 6.14. *Let $E : \mathcal{F} \to \mathcal{B}(\mathcal{H})$ be a spectral measure, as in Definition 6.8. For each h and g in $\mathcal{H}$, define*

$$\mu_{h,g}(S) = \langle E(S) h, g \rangle \tag{6.6}$$

for S in $\mathcal{F}$. This is a finite complex measure on $(X, \mathcal{F})$ and it has total variation $\|\mu_{h,g}\|$ at most $\|h\| \|g\|$. The measure is positive if $h = g$.

Proof. We first show that $\mu_{h,g}$ is countably additive. Suppose that $\{S_k\}$ is a sequence of pairwise disjoint sets in $\mathcal{F}$, and let S denote their union. We have

$$\mu_{h,g}(S) = \langle E(S)h, g \rangle = \langle \sum_1^\infty E(S_k)h, g \rangle$$

$$= \sum_1^\infty \langle E(S_k)h, g \rangle = \sum_1^\infty \mu_{h,g}(S_k).$$

When $h = g$ the positivity statement is a consequence of the fact that $E(S)$ is a self-adjoint idempotent:

$$\mu_{h,h}(S) = \langle E(S)h, h \rangle = \langle E(S)h, E(S)h \rangle \geq 0.$$

To verify the statement about the total variation recall what must be done: Given any partition of X into disjoint measurable subsets $\{S_k\}_{k=1}^\infty$, we must show that

$$\sum_1^\infty |\mu_{h,g}(S_k)| \leq \|h\| \, \|g\|.$$

Now for such a partition,

$$\sum_1^\infty |\mu_{h,g}(S_k)| = \sum_1^\infty |\langle E(S_k)h, g \rangle|$$

$$= \sum_1^\infty |\langle E(S_k)h, E(S_k)g \rangle|$$

$$\leq \sum_1^\infty \|E(S_k)h\| \, \|E(S_k)g\|$$

$$\leq \left(\sum_1^\infty \|E(S_k)h\|^2 \right)^{1/2} \left(\sum_1^\infty \|E(S_k)g\|^2 \right)^{1/2}$$

$$= \|E(X)h\| \, \|E(X)g\| = \|h\| \, \|g\|,$$

where we have used the Cauchy–Schwarz inequality in both $\mathcal{H}$ and ℓ^2, and Equation (6.5).   □

In the case of spectral measures $E : \mathcal{F} \to \mathcal{B}(\mathcal{H})$, where $\mathcal{F}$ denotes the Borel subsets of a compact set X in $\mathbb{C}$, the measures $\mu_{h,g}$ of Proposition 6.14 are necessarily regular measures (meaning the positive measures $|\mu_{h,g}|$ are regular). This follows, for example, by Theorem 2.18 in [40].

Given a spectral measure $E : \mathcal{F} \to \mathcal{B}(\mathcal{H})$, we want to define an operator which deserves to be called

$$\int f \, dE,$$

where f is a bounded measurable complex function on $(X, \mathcal{F})$. We discuss this heuristically first. It seems reasonable to assign to the choice $f = \chi_S$ the projection $E(S)$, so that we would have

$$\int \chi_S \, dE = E(S).$$

Certainly we would want our "integration" to be linear, so when f is a measurable simple function, that is

$$f = \sum_{k=1}^{n} a_k \chi_{S_k},$$

our definition should yield

$$\int f \, dE = \sum_{k=1}^{n} a_k E(S_k).$$

Notice that with such a choice for f, we would have

$$\left\langle \left(\int f dE \right) h, g \right\rangle = \langle \sum_{k=1}^{n} a_k E(S_k) h, g \rangle = \sum_{k=1}^{n} a_k \langle E(S_k) h, g \rangle$$

$$= \sum_{k=1}^{n} a_k \mu_{h,g}(S_k)$$

$$= \int \left(\sum_{k=1}^{n} a_k \chi_{S_k} \right) d\mu_{h,g}$$

$$= \int f d\mu_{h,g}$$

for all h and g in $\mathscr{H}$.

This suggests how we should proceed to formally define $\int f dE$, for any bounded measurable f, which we do next. Fix such a function f and consider $b : \mathscr{H} \times \mathscr{H} \to \mathbb{C}$ defined by

$$b(h,g) \equiv \int f d\mu_{h,g}.$$

Since $\mu_{h,g}$ is a complex measure, we remind the reader how the integral on the right hand side of this equation is defined. As a consequence of the Radon–Nikodym theorem, for a given complex measure v on a set X there exists a measurable function $m(x)$ with modulus 1 so that $dv = m d|v|$, where $|v|$ is the total variation measure of v, and integration with respect to v is defined to be integration with respect to $m d|v|$. Thus

$$\int f d\mu_{h,g} = \int fm \, d|\mu_{h,g}|,$$

and m is the Radon–Nikodym derivative of $\mu_{h,g}$ with respect to $|\mu_{h,g}|$. Since for each measurable set S, $\mu_{h,g}(S) = \langle E(S)h, g \rangle$, we see that b is a sesquilinear form on $\mathscr{H} \times \mathscr{H}$. Moreover, by Proposition 6.14,

$$|b(h,g)| \le \int |f| \, d|\mu_{h,g}| \le \|f\|_X \|\mu_{h,g}\| \le \|f\|_X \|h\|_{\mathscr{H}} \|g\|_{\mathscr{H}},$$

where $\|f\|_X$ denotes the supremum norm of f on X, and so b is bounded. By Theorem 2.11 there exists a unique bounded operator $\pi(f)$ on $\mathscr{H}$ with

$$b(h,g) = \langle \pi(f)h, g \rangle \qquad (6.7)$$

for all g and h in $\mathscr{H}$, and moreover

$$\|\pi(f)\| \le \|f\|_X. \qquad (6.8)$$

Since we have

$$\langle \pi(f)h, g \rangle = \int f d\mu_{h,g}$$

for all h and g, we define $\int f dE$ to be the operator $\pi(f)$. In short

$$A = \int f \, dE$$

is the unique operator with

$$\langle Ah, g \rangle = \int f d\mu_{h,g}$$

for all h, g in $\mathscr{H}$.

It follows that

$$\int \chi_S \, dE = \pi(\chi_S) = E(S)$$

for any set $S \in \mathscr{F}$, as desired in our heuristic discussion above. More generally, using the linearity of π as verified in Theorem 6.15 below, we see that for a simple function

$$u = \sum_{k=1}^{n} a_k \chi_{S_k}$$

we have

$$\int u \, dE = \pi(u) = \sum_{k=1}^{n} a_k E(S_k).$$

Thus $\pi(u)$ is a linear combination of projections when u is simple.

If the sets S_k are pairwise disjoint, $\pi(u)$ has a particularly nice geometric form. Disjointness guarantees that the subspaces $E(S_k)\mathscr{H}$ and $E(S_m)\mathscr{H}$ are orthogonal for $k \ne m$. Each subspace $E(S_k)\mathscr{H}$ is invariant for the operator $\pi(u)$, and on this subspace $\pi(u)$ acts as the scalar a_k times the identity. Writing $\mathscr{H}$ as the orthogonal direct sum

$$\mathscr{H} = \mathscr{K} \oplus \sum_{k=1}^{n} \oplus E(S_k)\mathscr{H}, \qquad (6.9)$$

where $\mathscr{K}$ is defined by this equation, we observe that $\pi(u)$ acts on $\mathscr{K}$ as the zero operator. Thus with regard to the decomposition (6.9) of $\mathscr{H}$, $\pi(u)$ acts as a block diagonal matrix

$$\begin{bmatrix} 0_{\mathscr{K}} & & & 0 \\ & a_1 I_1 & & \\ & & \ddots & \\ 0 & & & a_n I_n \end{bmatrix}$$

where I_k is the identity on $E(S_k)\mathscr{H}$ (see Exercise 4.18 in Chapter 4 for the notion of a matrix with operator entries).

It is worth noting that given any bounded $\mathscr{F}$-measurable function f on X, $\pi(f)$ may be approximated in operator norm by operators of the form $\pi(u)$ with u a simple function as above. In view of the inequality (6.8) (and the linearity of π established below in the proof of Theorem 6.15), it's enough to produce from f and an arbitrary $\varepsilon > 0$ a simple function u with $\|f - u\|_X < \varepsilon$. The standard strategy for doing this (variants of which are implemented in the proofs of Theorems 6.15 and 6.21 below) is to choose pairwise disjoint Borel measurable sets $R_1, R_2, \ldots, R_q$ in $\mathbb{C}$, having diameters less than ε, each intersecting the range of f, whose union contains the range of f. The sets $A_k \equiv f^{-1}(R_k)$, $k = 1, 2, \ldots, q$, will form a partition of X into nonempty pairwise disjoint measurable sets, and the simple function

$$u \equiv \sum_{k=1}^{q} f(x_k)\chi_{A_k},$$

where x_k is an arbitrary point in A_k, does the job:

$$\|\pi(f) - \pi(u)\| = \|\pi(f - u)\| \le \|f - u\|_X < \varepsilon.$$

In the next result, we denote the C^*-algebra of bounded $\mathscr{F}$-measurable functions on X, in the supremum norm, with pointwise operations and involution $f^* = \bar{f}$, by $B(X, \mathscr{F})$.

Theorem 6.15. *The map $\pi : B(X, \mathscr{F}) \to \mathscr{B}(\mathscr{H})$ is a $*$-homomorphism.*

Proof. We show first that π is linear. Let h and g be in $\mathscr{H}$ and suppose f_1 and f_2 are bounded measurable functions on X and α is a complex scalar. We have

$$\begin{aligned} \langle \pi(\alpha f_1 + f_2)h, g \rangle &= \int (\alpha f_1 + f_2) d\mu_{h,g} \\ &= \alpha \int f_1 d\mu_{h,g} + \int f_2 d\mu_{h,g} \\ &= \alpha \langle \pi(f_1)h, g \rangle + \langle \pi(f_2)h, g \rangle \\ &= \langle (\alpha\pi(f_1) + \pi(f_2))h, g \rangle. \end{aligned}$$

This shows that $\pi(\alpha f_1 + f_2) = \alpha\pi(f_1) + \pi(f_2)$ and thus π is linear. The fact that $\pi(1) = I$ follows from the computation

$$\langle \pi(1)h, g \rangle = \mu_{h,g}(X) = \langle Ih, g \rangle.$$

Furthermore, for any bounded measurable f,

$$\langle \pi(\overline{f})h, g \rangle = \int \overline{f} d\mu_{h,g} = \overline{\int f d\overline{\mu_{h,g}}}.$$

But $\overline{\mu_{h,g}}(S) = \overline{\langle g, E(S)h \rangle} = \mu_{g,h}(S)$, so that

$$\langle \pi(\overline{f})h, g \rangle = \overline{\langle \pi(f)g, h \rangle} = \overline{\langle g, \pi(f)^*h \rangle} = \langle \pi(f)^*h, g \rangle,$$

which says that $\pi(\overline{f}) = \pi(f)^*$.

It remains to show that π is multiplicative. Let $\varepsilon > 0$ and partition the plane $\mathbb{C}$ (using equally spaced horizontal and vertical lines) into a grid of pairwise disjoint, equal-size semiclosed squares of diameter less than ε (to be specific, say that a square includes its west and south edges, but not its east and north edges, and of the corners, only includes the southwest corner). Since our functions f_1 and f_2 are bounded, only finitely many of these squares will intersect either the range of f_1 or the range of f_2; label them $R_1, R_2, \ldots R_q$, and note that their union contains the ranges of both f_1 and f_2. Set $A_k = f_1^{-1}(R_k)$ and $B_k = f_2^{-1}(R_k)$ for each k, $1 \le k \le q$, and from the sets $A_i \cap B_j$, $1 \le i, j \le q$, list only those that are nonempty, denoting them $C_1, C_2, \ldots, C_p$. This gives a partition of X into pairwise disjoint pieces. If x, y lie in the same piece C_k, then $|f_1(x) - f_1(y)| < \varepsilon$ and $|f_2(x) - f_2(y)| < \varepsilon$. Choose a sampling point x_k in each C_k and define the simple functions

$$u = \sum_{k=1}^{p} f_1(x_k)\chi_{C_k}$$

and

$$v = \sum_{k=1}^{p} f_2(x_k)\chi_{C_k}.$$

Note that $\|f_1 - u\|_X < \varepsilon$ and $\|f_2 - v\|_X < \varepsilon$. Since the sets C_k are pairwise disjoint,

$$uv = \sum_{k=1}^{p} f_1(x_k)f_2(x_k)\chi_{C_k}.$$

We may easily compute the product $\pi(u)\pi(v)$. Since $E(C_j)\mathscr{H} \perp E(C_k)\mathscr{H}$ if $j \ne k$, we have

$$\pi(u)\pi(v) = \left(\sum_{j=1}^{p} f_1(x_j)E(C_j) \right) \left(\sum_{k=1}^{p} f_2(x_k)E(C_k) \right)$$

$$= \sum_{j,k=1}^{p} f_1(x_j) f_2(x_k) E(C_j) E(C_k)$$

$$= \sum_{k=1}^{p} f_1(x_k) f_2(x_k) E(C_k)$$

$$= \pi(uv).$$

Replace ε by a sequence $\varepsilon_n \to 0$ and u and v by corresponding sequences u_n and v_n. Note that u_n, v_n, and $u_n v_n$ converge uniformly to f_1, f_2, and $f_1 f_2$, respectively. Thus by (6.8)

$$\pi(f_1 f_2) = \lim_{n \to \infty} \pi(u_n v_n) = \lim_{n \to \infty} \pi(u_n) \pi(v_n) = \pi(f_1) \pi(f_2)$$

as desired.   □

Corollary 6.16. *For each $f \in B(X, \mathscr{F})$, $\pi(f)$ is a normal operator in $\mathscr{B}(\mathscr{H})$.*

Proof. We have

$$\pi(f)\pi(f)^* = \pi(f)\pi(\overline{f}) = \pi(f\overline{f}) = \pi(f)^* \pi(f).$$

□

The properties of π given in Theorem 6.15 can be written in spectral integral notation as

$$\int (\alpha f_1 + f_2)\, dE = \alpha \int f_1\, dE + \int f_2\, dE,$$

$$\int \overline{f}\, dE = \left(\int f\, dE \right)^*,$$

and

$$\int f_1 f_2\, dE = \left(\int f_1\, dE \right) \left(\int f_2\, dE \right)$$

for bounded $\mathscr{F}$-measurable functions f, f_1, and f_2 and scalar α.

By a *representation* of a unital C^*-algebra, we mean a $*$-homomorphism of the algebra into $\mathscr{B}(\mathscr{H})$ for some Hilbert space $\mathscr{H}$ which maps I to I. Theorem 6.15 shows that π is a representation of $B(X, \mathscr{F})$. A representation is termed *faithful* if it is injective, and by Exercise 5.19 of Chapter 5, any faithful representation is an isometry. Although we do not prove it here, it turns out that *every* C^*-algebra has a representation which is an isometry. This says that every C^*-algebra "is" a subalgebra of $\mathscr{B}(\mathscr{H})$ for some Hilbert space $\mathscr{H}$.

We next explore some consequences of the properties of the representation π in Theorem 6.15. If $(X, \mathscr{F})$ is a measure space and $E : \mathscr{F} \to \mathscr{B}(\mathscr{H})$ is a spectral measure we say that a set $C \in \mathscr{F}$ is a *carrier* for E if $E(X \setminus C) = 0$. Since $X = C \cup (X \setminus C)$ and $I = E(C) + E(X \setminus C)$, C is a carrier exactly when $\pi(\chi_C) = E(C) = I$. In Exercise 6.9 you are asked to show that if f_1 and f_2 are bounded measurable functions on X with $f_1(x) = f_2(x)$ for all x in some carrier C for E, then

$$\int f_1 \, dE = \int f_2 \, dE.$$

As an application of these ideas, suppose $X = \mathbb{C}$ and E is a spectral measure on the Borel sets of $\mathbb{C}$. The identity function z is of course not bounded on $\mathbb{C}$, but if E has a bounded carrier C, so that $z\chi_C$ is a bounded measurable function on $\mathbb{C}$, then we may define

$$\int z \, dE \equiv \int z\chi_C \, dE.$$

We claim that the right-hand side is independent of the choice of bounded carrier C for E. If C_1 and C_2 are both bounded carriers, then so is $C_1 \cap C_2$. This observation follows from noting that

$$\mathbb{C}\backslash(C_1 \cap C_2) = (\mathbb{C}\backslash C_1) \cup (\mathbb{C}\backslash C_2) = (\mathbb{C}\backslash C_1) \cup (C_1\backslash C_2)$$

and that $\mathbb{C}\backslash C_1$ and $C_1\backslash C_2$ are disjoint with $E(\mathbb{C}\backslash C_1) = 0$ and $E(C_1\backslash C_2) \leq E(\mathbb{C}\backslash C_2) = 0$. Since $z\chi_{C_1} = z\chi_{C_2}$ on the carrier $C_1 \cap C_2$, we apply Exercise 6.9 to conclude that

$$\int z\chi_{C_1} \, dE = \int z\chi_{C_2} \, dE.$$

Recall from our discussion following Example 6.11 that the support of spectral measure $E : \mathscr{F} \to \mathscr{B}(\mathscr{H})$ defined on the Borel sets $\mathscr{F}$ of $\mathbb{C}$ is the complement of the union of all open sets U with $E(U) = 0$. The support K is a carrier of E. Although carriers of E are not unique and need not be closed, the support K is unique (and of course also closed).

Proposition 6.17. *Let E be a spectral measure defined on the Borel subsets of $\mathbb{C}$. The following are equivalent:*

(a) E has a bounded carrier.
(b)   The support of E is compact.

The proof is left to the reader as Exercise 6.10.

The proof of the next result, which follows easily from the properties of π, is also left to the reader, as Exercise 6.11

Proposition 6.18. *Let $(X, \mathscr{F})$ be a measure space and $E : \mathscr{F} \to \mathscr{B}(\mathscr{H})$ a spectral measure. If f is a bounded measurable function on X and h is in $\mathscr{H}$, then*

$$\left\| \left(\int f \, dE \right) h \right\|^2 = \int |f|^2 \, d\mu_{h,h}.$$

We use Proposition 6.18 in the next result.

Proposition 6.19. *Let $(X, \mathscr{F}), E,$ and f be as in Proposition 6.18. We have*

$$\ker \left(\int f \, dE \right) = E(S_0)\mathscr{H},$$

where

$$S_0 = \{x \in X : f(x) = 0\}.$$

Proof. Write $\pi(f) = \int f\, dE$. By Proposition 6.18,

$$\|\pi(f)h\|^2 = \int |f|^2 \, d\mu_{h,h}$$

for any $h \in \mathcal{H}$. Thus

$$h \in \ker \pi(f) \iff \int |f|^2 \, d\mu_{h,h} = 0 \iff \mu_{h,h}(X \setminus S_0) = 0.$$

But

$$
\begin{aligned}
\|E(X \setminus S_0)h\|^2 &= \langle E(X \setminus S_0)h, E(X \setminus S_0)h \rangle \\
&= \langle E(X \setminus S_0)h, h \rangle \\
&= \mu_{h,h}(X \setminus S_0),
\end{aligned}
$$

so $h \in \ker \pi(f)$ if and only if $\|E(X \setminus S_0)h\| = 0$. Since $I = E(X) = E(X \setminus S_0) + E(S_0)$ where the two summands on the right-hand side have orthogonal ranges,

$$\|h\|^2 = \|E(S_0)h\|^2 + \|E(X \setminus S_0)h\|^2$$

and

$$h \in \ker \pi(f) \iff \|E(S_0)h\|^2 = \|h\|^2 \iff E(S_0)h = h \iff h \in E(S_0)\mathcal{H}.$$

$\square$

In the next result we investigate the eigenvalues of $\int z\, dE$.

Proposition 6.20. *Let $\mathcal{F}$ be the Borel subsets of $\mathbb{C}$ and suppose $E : \mathcal{F} \to \mathcal{B}(\mathcal{H})$ is a spectral measure with compact support. Set $A = \int z\, dE$ and suppose $\lambda_0 \in \mathbb{C}$. We have*

$$E(\{\lambda_0\}) \neq 0 \iff \lambda_0 \text{ is an eigenvalue of } A.$$

In this case,

$$\ker (A - \lambda_0 I) = E(\{\lambda_0\})\mathcal{H}.$$

Proof. Let D be a closed disk in $\mathbb{C}$ containing both λ_0 and the support of E. Clearly D is a carrier of E. If we set $f = z\chi_D$,

$$A = \int f\, dE = \pi(f).$$

By Proposition 6.19,

$$\ker (A - \lambda_0 I) = \ker \pi(f - \lambda_0) = E(S_0)\mathcal{H},$$

where $S_0 = \{\lambda \in \mathbb{C} : f(\lambda) = \lambda_0\}$. Since

$$E(S_0) = E(S_0 \backslash D) + E(S_0 \cap D) = 0 + E(\{\lambda_0\})$$

we see that ker $(A - \lambda_0 I) = E(\{\lambda_0\})\mathscr{H}$ and in particular, λ_0 is an eigenvalue of A if and only if $E(\{\lambda_0\}) \neq 0$. $\square$

The next result provides the key idea for the spectral measure version of the spectral theorem. It uses the spectral measure of Example 6.11, associated to a multiplication operator.

Theorem 6.21. *Let (X, μ) be a σ-finite measure space and let φ be in $L^\infty(X, \mu)$. Consider the operator M_φ acting on $L^2(X, \mu)$ and the spectral measure*

$$E(S) = M_{\chi_{\varphi^{-1}(S)}}$$

for S a Borel set of $\mathbb{C}$. Then E has compact support, and

$$M_\varphi = \int z \, dE.$$

Proof. By Exercise 6.5, the support of E coincides with the essential range of φ, that is, with $\sigma(M_\varphi)$, which is compact. Choose a representative of φ whose range is contained in the essential range of φ. Let $\varepsilon > 0$ be arbitrary. We begin by constructing a finite collection of pairwise disjoint Borel sets in $\mathbb{C}$, each having diameter less than ε and intersecting the range of φ, and whose union is a closed set containing the range of φ. Any such collection of sets will serve our purposes here, and we indicate one possible procedure for carrying out such a construction. Take a closed square in $\mathbb{C}$ which contains the range of φ. Cover this square with a grid of closed sub-squares each of diameter less than ε. Let $C_1, C_2, \ldots, C_q$ be those closed sub-squares which intersect the range of φ. Set $D_1 = C_1$. Find the smallest value of $k \leq q$, call it k_1, such that $C_{k_1} \backslash D_1$ intersects the range of φ and set

$$D_2 = C_{k_1} \backslash D_1.$$

If no such k_1 exists, the process terminates with D_1. If k_1, and hence D_2, do exist, notice that $D_1 \cap D_2 = \emptyset$ and $D_1 \cup D_2 = C_1 \cup C_{k_1}$. Next, find the smallest $k_2 > k_1$ so that $C_{k_2} \backslash (D_1 \cup D_2)$ intersects the range of φ, and define

$$D_3 = C_{k_2} \backslash (D_1 \cup D_2).$$

If no such k_2 exists, the process terminates with D_2. If k_2 (and thus D_3) exist, the sets D_1, D_2, D_3 are pairwise disjoint and

$$D_1 \cup D_2 \cup D_3 = C_1 \cup C_{k_1} \cup C_{k_2}.$$

Continue, so that if sets $D_1, D_2, \ldots, D_m$ have been constructed, at the next step we seek the least k_m with $q \geq k_m > k_{m-1}$ so that $C_{k_m} \backslash (D_1 \cup \cdots \cup D_m)$ intersects the range of φ; if no such k_m exists, the process terminates with D_m. Otherwise, the set D_{m+1} is defined by

$$D_{m+1} = C_{k_m} \setminus (D_1 \cup \cdots \cup D_m).$$

Eventually the process must terminate, and the resulting collection $D_1, D_2, \ldots, D_p$ are pairwise disjoint Borel sets, each having diameter less than ε, with

$$\text{range } \varphi \subseteq \cup_{k=1}^p D_k \equiv K.$$

Since

$$K = D_1 \cup D_2 \cdots \cup D_p = C_1 \cup C_{k_1} \cup \cdots \cup C_{k_{p-1}}$$

is a union of finitely many closed squares from our original collection of C_k, K is a closed set.

Define

$$A_k \equiv \varphi^{-1}(D_k)$$

for $k = 1, 2, \ldots, p$ and note that $A_1, A_2, \ldots, A_p$ forms a partition of X into pairwise disjoint, nonempty, μ-measurable sets. Select a sampling point x_k in A_k for each k. The simple function

$$u = \sum_{k=1}^p \varphi(x_k) \chi_{D_k} \tag{6.10}$$

is a good approximation to the identity function on K:

$$\|u - z\|_K \equiv \sup_{z \in K} |u(z) - z| \le \varepsilon. \tag{6.11}$$

Write π for the representation associated to E, so that $\pi(f) = \int f \, dE$. We have

$$\pi(u) = \sum_{k=1}^p \varphi(x_k) E(D_k)$$

and

$$E(D_k) = M_{\chi_{\varphi^{-1}(D_k)}} = M_{\chi_{A_k}}.$$

Since the A_k form a partition of X,

$$\sum_{k=1}^p \chi_{A_k} = 1.$$

We see that for any f in $L^2(X, \mu)$

$$\|(M_\varphi - \pi(u)) f\|^2 = \left\| \varphi \left(\sum_{k=1}^p \chi_{A_k} \right) f - \left(\sum_{k=1}^p \varphi(x_k) E(D_k) \right) f \right\|^2$$

$$= \left\| \sum_{k=1}^p (\varphi - \varphi(x_k)) \chi_{A_k} f \right\|^2.$$

Since the sets $A_1, A_2, \ldots, A_p$ partition X and $|\varphi - \varphi(x_k)| < \varepsilon$ on A_k, the above coincides with

$$\sum_{k=1}^{p} \int_{A_k} |\varphi - \varphi(x_k)|^2 |f|^2 d\mu,$$

which is bounded above by

$$\varepsilon^2 \sum_{k=1}^{p} \int_{A_k} |f|^2 d\mu = \varepsilon^2 \|f\|^2.$$

This says that

$$\|M_\varphi - \pi(u)\| \leq \varepsilon. \tag{6.12}$$

On the other hand, the inequality (6.8) applied to $f = u - z$, yields

$$\|\pi(u) - \pi(z)\| = \|\pi(u - z)\| \leq \|u - z\|_K \leq \varepsilon$$

since K, a closed set containing the range of φ, contains the support of E. Thus

$$\|M_\varphi - \pi(z)\| \leq 2\varepsilon,$$

and since ε was arbitrary we conclude that

$$M_\varphi = \pi(z) = \int z \, dE.$$

$\square$

We look next at unitarily equivalent spectral measures.

Proposition 6.22. *Let $\mathscr{H}$ and $\mathscr{K}$ be Hilbert spaces and suppose that $(X, \mathscr{F})$ is a measure space. Let $E : \mathscr{F} \to \mathscr{B}(\mathscr{H})$ and $F : \mathscr{F} \to \mathscr{B}(\mathscr{K})$ be spectral measures. If there is a unitary operator $W : \mathscr{H} \to \mathscr{K}$ such that $F(S) = WE(S)W^{-1}$ for all $S \in \mathscr{F}$, then*

$$\int f \, dF = W \left(\int f \, dE \right) W^{-1}$$

for all bounded $\mathscr{F}$-measurable functions f on X.

Proof. For any bounded measurable function f on X we let

$$\pi(f) = \int f \, dE \text{ and } \rho(f) = \int f \, dF.$$

Let $\varepsilon > 0$ and choose, as we have done above, a simple function u on X with

$$\|f - u\|_X < \varepsilon.$$

If

$$u = \sum_{k=1}^{q} c_k \chi_{A_k},$$

then

$$\pi(u) = \sum_{k=1}^{q} c_k E(A_k) \text{ and } \rho(u) = \sum_{k=1}^{q} c_k F(A_k)$$

so that

$$W\pi(u)W^{-1} = \sum_{k=1}^{q} c_k WE(A_k)W^{-1} = \sum_{k=1}^{q} c_k F(A_k) = \rho(u).$$

Thus

$$\begin{aligned}
\|W\pi(f)W^{-1} - \rho(f)\| &= \|W(\pi(f) - \pi(u))W^{-1} + \rho(u) - \rho(f)\| \\
&\le \|W(\pi(f) - \pi(u))W^{-1}\| + \|\rho(u) - \rho(f)\| \\
&= \|\pi(f - u)\| + \|\rho(f - u)\| \le 2\|f - u\|_X < 2\varepsilon.
\end{aligned}$$

Since ε is arbitrary, $\rho(f) = W\pi(f)W^{-1}$. $\qquad\square$

The next result is the "spectral measure version" of the spectral theorem. The reader is encouraged to see a concrete application of its statement by working out Exercises 6.13 and 6.14 before proceeding to the proof of the theorem. The intuition behind this theorem is that just as a bounded, Borel measurable function can be approximated by linear combinations of characteristic functions associated to pairwise disjoint sets (i.e., simple functions), bounded normal operators can be approximated by linear combinations of projections with pairwise orthogonal ranges.

Theorem 6.23 (Spectral Theorem, Spectral Measure Version). *Let A be a normal operator in $\mathscr{B}(\mathscr{H})$, where $\mathscr{H}$ is a separable Hilbert space. Let $\mathscr{F}$ denote the Borel subsets of $\sigma(A)$. There is a unique spectral measure $F : \mathscr{F} \to \mathscr{B}(\mathscr{H})$ with*

$$A = \int z\, dF.$$

Moreover, given any continuous function f on $\sigma(A)$,

$$f(A) = \int f\, dF,$$

where on the left-hand side $f(A)$ is defined by the continuous functional calculus (Theorem 5.46).

Proof. We address the existence part of the statement first. By Theorem 6.2 we know that there is a σ-finite measure space (X, v), a unitary operator $W : L^2(X, v) \to \mathscr{H}$, and a $\varphi \in L^\infty(X, v)$ so that $A = WM_\varphi W^{-1}$. Let E be the spectral measure associated to M_φ as in Example 6.11. We know from Theorem 6.21 that

$$\int z\, dE = M_\varphi.$$

By Example 6.12, $F(S) \equiv WE(S)W^{-1}$ is a also spectral measure defined on the Borel subsets of $\sigma(A)$. We claim that $A = \int z\, dF$. This is immediate from Proposition 6.22, with f taken to be the restriction of z to $\sigma(A)$. Indeed,

$$\int z\,dF = W\left(\int z\,dE\right)W^{-1} = WM_\varphi W^{-1} = A.$$

This proves the existence part of the first statement of the theorem; the proof of the uniqueness statement is outlined in Exercise 6.16.

For the second statement of the theorem, let f be continuous on $\sigma(A)$ and set $\pi(f) = \int f\,dF$. By Theorem 6.15, $p(A,A^*) = \int p(z,\bar{z})\,dF$ for any polynomial p in z and $\bar{z}$. Since f is a uniform limit on $\sigma(A)$ of a sequence of such polynomials, we have

$$f(A) = \int f\,dF$$

as desired.   □

For a normal operator A in $\mathcal{B}(\mathcal{H})$, the expression "the spectral measure of A" always refers to the unique measure given in Theorem 6.23. Notice that when E is the spectral measure of A, the statement

$$g(A) = \int g\,dE$$

holds for any function g which is continuous on $\sigma(A)$. The same identity, with "$g(A)$" defined by the Borel functional calculus (that is, by Equation (6.3)), holds for any bounded Borel measurable function; see Exercise 6.18.

By Proposition 6.20 we know that when $\mathcal{H}$ is separable and A is a normal operator in $\mathcal{B}(\mathcal{H})$ with spectral measure E, then the eigenvalues of A are precisely those points $\lambda_0 \in \sigma(A)$ for which $E(\{\lambda_0\})$ is nonzero, and moreover $E(\{\lambda_0\})\mathcal{H} =$ ker $(A - \lambda_0 I)$. Let's interpret this in the context of a *compact* normal operator T. We know from Theorem 4.31 that in this case the nonzero points of $\sigma(T)$ are all eigenvalues, and this set is at most countable, accumulating only at zero if infinite. Denote these nonzero eigenvalues by $\lambda_1, \lambda_2, \ldots$. Set $E_k = E(\{\lambda_k\})$, where E is the spectral measure of T. Each E_k is the projection onto ker$(T - \lambda_k I)$. Moreover, by Exercise 4.10 in Chapter 4, E_k is finite rank (that is, it is the projection onto a finite-dimensional subspace), since a compact operator has finite-dimensional eigenspaces corresponding to its nonzero eigenvalues, and $E_k\mathcal{H} \perp E_n\mathcal{H}$ for $k \neq n$.

What is $\sum_k \lambda_k E_k$? Set $f_k = \lambda_k \chi_{\{\lambda_k\}}$ and notice that

$$\sum_k f_k(z) = z$$

on $\sigma(T)$ and that $f_k(T) = \lambda_k E_k$. Thus

$$T = \sum_k f_k(T) = \sum \lambda_k E_k.$$

If the set of nonzero eigenvalues λ_k is infinite, the sum converges in the norm of $\mathcal{B}(\mathcal{H})$, since compactness of T implies that $\lambda_k \to 0$ as $k \to \infty$. Notice that with this application of Theorem 6.23, we have recovered Theorem 4.24, in the case that T

is compact and self-adjoint, and provided an extension of Theorem 4.24 to compact normal operators.

6.3 Exercises

6.1. Suppose a normal operator A in $\mathcal{B}(\mathcal{H})$ has enough eigenvectors to provide an orthonormal basis for $\mathcal{H}$:

$$Ae_k = \alpha_k e_k$$

where $\{e_k\}$ is an orthonormal basis for $\mathcal{H}$.

(a) Check that the sequence $(\alpha_1, \alpha_2, \alpha_3, \ldots)$ is in $\ell^\infty(\mathbb{N})$.
(b) Define $W : \ell^2 \to \mathcal{H}$ by $W(\lambda_1, \lambda_2, \ldots) = \lambda_1 e_1 + \lambda_2 e_2 + \cdots$. Check that W is a unitary map of ℓ^2 onto $\mathcal{H}$.
(c) Define the operator B in $\mathcal{B}(\ell^2)$ by $B \equiv W^{-1}AW$. Identify B as a multiplication operator on ℓ^2, where we regard $\ell^2 = L^2(\mathbb{N})$ in the usual way.

6.2. Let $\mathcal{H} = \mathbb{C}^n$ and let A be the operator on $\mathcal{H}$ given by the $n \times n$ diagonal matrix with diagonal $(a_1, a_2, \ldots, a_n)$.

Show that A has a cyclic vector if and only if all the diagonal entries are distinct. Hints: If the a_j are distinct, consider the vector h in $\mathbb{C}^n$ of all 1's. For the converse, mimic the argument in Example 6.4.

6.3. Show that every normal operator A acting on a separable Hilbert space is unitarily equivalent to a multiplication operator on $L^2(X, \mu)$ for some *finite* measure space (X, μ). Hint: If $\mathcal{H}_n$ is invariant under $C^*(A)$ and has cyclic vector h_n, then we may assume $\|h_n\| = 2^{-n}$.

6.4. A *conjugation* on a Hilbert space $\mathcal{H}$ is a conjugate linear map $C : \mathcal{H} \to \mathcal{H}$ with $C^2 = I$ and $\langle Cx, Cy \rangle = \langle y, x \rangle$ for all x, y in $\mathcal{H}$.

(a) Show that if we fix an orthonormal basis $\{e_n\}$ for $\mathcal{H}$ and set

$$C\left(\sum \lambda_n e_n\right) = \sum \overline{\lambda_n} e_n$$

(provided $\{\lambda_n\} \in \ell^2$) then C is a conjugation.
(b) If $\mathcal{H} = L^2(X, \mu)$ for some σ-finite measure space (X, μ), show that $C(f) = \overline{f}$ is a conjugation.
(c) If C is a conjugation, an operator $T \in \mathcal{B}(\mathcal{H})$ is called C-symmetric if $CT^* = TC$. Show that the Volterra operator $Vf(x) = \int_0^x f(t)dt$ on $L^2[0, 1]$ is C-symmetric for $Cf(x) = \overline{f(1-x)}$.
(d) If N is any normal operator in $\mathcal{B}(\mathcal{H})$, find a conjugation C on $\mathcal{H}$ so that $CN^* = NC$.

6.5. Verify the assertion made in Example 6.11 about the support of E.

6.6. Provide the details in Example 6.12.

6.7. Let E be a spectral measure on $(X, \mathscr{F})$. Show that if S_1 and S_2 are in $\mathscr{F}$, then $E(S_1 \cap S_2) = E(S_1)E(S_2)$.

6.8. Suppose that X is a topological space and $\mathscr{F}$ is the Borel σ-algebra on X. A spectral measure E on $(X, \mathscr{F})$ is said to be *regular* if

$$E(S)\mathscr{H} = \text{closed linear span } \{E(K)\mathscr{H} : K \subseteq S \text{ is compact}\}$$

for every S in $\mathscr{F}$. Show that if $X = \mathbb{C}$, then any spectral measure $E : \mathscr{F} \to \mathscr{B}(\mathscr{H})$ is necessarily regular. Hints: Clearly,

$$\text{closed linear span } \{E(K)\mathscr{H} : K \subseteq S \text{ is compact}\} \subseteq E(S)\mathscr{H}.$$

For the reverse inclusion, suppose that h is a nonzero vector in $\mathscr{H}$ with $h \perp E(K)\mathscr{H}$ for all compact subsets K of S. Use the fact that the measure $\mu_{h,h}$ is automatically regular ([40]) to show that $h \perp E(S)\mathscr{H}$.

6.9. Suppose that $(X, \mathscr{F})$ is a measure space and $E : \mathscr{F} \to \mathscr{B}(\mathscr{H})$ is a spectral measure. Show that if f_1 and f_2 are bounded measurable functions on X with $f_1(x) = f_2(x)$ for all x in some carrier C for E, then

$$\int f_1 \, dE = \int f_2 \, dE.$$

6.10. Prove Proposition 6.17.

6.11. Prove Proposition 6.18.

6.12. Let $\mathscr{F}$ be the Borel subsets of $\mathbb{C}$ and suppose $E : \mathscr{F} \to \mathscr{B}(\mathscr{H})$ is a spectral measure with compact support K. Set $A = \int z \, dE$. Show that $\sigma(A) = K$. Hint: If $\lambda_0 \in K$, show that $A - \lambda_0 I$ is not bounded below by considering unit vectors in $E(D_\varepsilon(\lambda_0))\mathscr{H}$, where $D_\varepsilon(\lambda_0)$ is the open disk of radius ε centered at λ_0.

6.13. Let T be the diagonal operator with diagonal $\{\lambda_1, \lambda_2, \ldots\}$ on ℓ^2. Consider the spectral measure E defined in Example 6.10. Describe concretely the associated measures $\mu_{h,g}$, the operator $\pi(z)$, and verify directly that

$$T = \int z \, dE.$$

6.14. Let A be the normal operator of multiplication by $\varphi(x) = x$ on $L^2[0, 1]$. Let $E : [0, 1] \to \mathscr{B}(L^2[0, 1])$ be the spectral measure of A as in Theorem 6.23. Identify E.

6.15. Suppose E is a spectral measure and suppose that $E(S)$ commutes with an operator T for each S in $\mathscr{F}$. Show that for any $f \in B(X, \mathscr{F})$, $\int f \, dE$ commutes with T.

6.16. Prove the uniqueness statement in Theorem 6.23. One possible outline for the argument is as follows.

(a) Your goal is to show that if $F : \mathscr{F} \to \mathscr{B}(\mathscr{H})$ is the spectral measure constructed in the proof of Theorem 6.23, and $F_1 : \mathscr{F} \to \mathscr{B}(\mathscr{H})$ is a second spectral measure with

$$\int z \, dF = \int z \, dF_1$$

then $F = F_1$. Argue that it suffices to show $\mu^1_{h,g} = \mu_{h,g}$ for all $h, g \in \mathscr{H}$, where $\mu^1_{h,g}(S) = \langle F_1(S)h, g \rangle$ and similarly for $\mu_{h,g}$.
(b) By polarization, argue that it suffices to show $\mu^1_{h,h} = \mu_{h,h}$ for all $h \in \mathscr{H}$.
(c) From $\int z \, dF = \int z \, dF_1$ we obtain $\int z \, d\mu_{h,h} = \int z \, d\mu^1_{h,h}$, as well as the analogous conclusion with $\bar{z}$ replacing z. We have

$$\int p(z, \bar{z}) \, d\mu_{h,h} = \int p(z, \bar{z}) \, d\mu^1_{h,h}$$

for polynomials p. Now invoke the Stone–Weierstrass theorem.

6.17. Suppose A is a normal operator in $\mathscr{B}(\mathscr{H})$ with spectrum X and let $E : \mathscr{F} \to \mathscr{B}(\mathscr{H})$ be its spectral measure as in Theorem 6.23.

(a) If X consists of a single point, show that A is a scalar multiple of the identity. Conclude that every subspace of $\mathscr{H}$ is an invariant subspace of A in this case.
(b) Show that if $X = S_1 \cup S_2$ where S_1 and S_2 are disjoint nonempty Borel subsets of X, then $E(S_1)$ commutes with A. Moreover, the ranges of $E(S_1)$ and $E(S_2)$ are invariant subspaces of A with $(E(S_1)\mathscr{H})^\perp = E(S_2)\mathscr{H}$.

6.18. For T any normal operator in $\mathscr{B}(\mathscr{H})$ and g any bounded Borel measurable function on $X = \sigma(T)$, we define $g(T)$ by the Borel functional calculus (Equation (6.3)). Show that $g(T) = \int g \, dE$ where E is the unique spectral measure in Theorem 6.23.

6.19. Suppose that A is a normal operator in $\mathscr{B}(\mathscr{H})$ and that λ_0 is an isolated point in $\sigma(A)$. Show that λ_0 is an eigenvalue of A.

6.20. Suppose that $\mathscr{H}$ is an infinite-dimensional separable Hilbert space and that $T \in \mathscr{B}(\mathscr{H})$ is a compact normal operator. Let f be any function in $C(\sigma(T))$. Show that $f(T)$ is compact if and only if $f(0) = 0$.

6.21. Suppose A is normal on $\mathscr{B}(\mathscr{H})$, so that $f(A)$ is also normal for any $f \in C(\sigma(A))$. If E_1 is the spectral measure of A and E_2 is the spectral measure of $f(A)$, show that $E_2(S) = E_1(f^{-1}(S))$ for all Borel sets S in $\sigma(f(A))$.

6.22. Let M and N be closed subspaces of a Hilbert space $\mathscr{H}$. Let P and Q be, respectively, the orthogonal projections of $\mathscr{H}$ onto M and N.

(a) Show that the sequence $\{A_n\}$ of operators given by

$$A_n = PQPQ \cdots PQP$$

(with n Q's and $(n+1)P$'s) converges in the strong operator topology to the orthogonal projection of $\mathscr{H}$ onto $M \cap N$. Hint: Apply the spectral measure version of the spectral theorem to the operator PQP.

(b) The result in part (a) has been called the zig-zag theorem. Draw a picture (as if $\mathscr{H}$ were the real Hilbert space $\mathbb{R}^3$ and M and N were two-dimensional subspaces) illustrating why.

Appendix A
Real Analysis Topics

As a way to illustrate the difference between the Riemann integral and the Lebesgue integral, consider the analogy of finding the value of a pile of coins. The "Riemann" way is to take the coins as they appear, adding the value of each piece as you pick it up. By contrast, in the Lebesgue method we start by sorting the coins by type — penny, nickel, dime, ... — and total the value as

$$1 \cdot m(A_1) + 5 \cdot m(A_2) + 10 \cdot m(A_3) + \cdots$$

where $m(A_1)$ is the number of pennies, $m(A_2)$ the number of nickels, and so on.[1] Notice how this approach suggests an interest in sums of the form $\sum s_j m(A_j)$ where "m" is some way of measuring the "size" of sets under consideration. Thus before we pursue a definition of the Lebesgue integral we will discuss the notion of measures.

A.1 Measures

Definition A.1. A measure space $(X, \mathfrak{M}, \mu)$ consists of a set X, a collection $\mathfrak{M}$ of subsets of X, and a function $\mu : \mathfrak{M} \to [0, \infty]$. The collection $\mathfrak{M}$ is a σ-algebra, that is, it is required to satisfy

(1) $X \in \mathfrak{M}$
(2) If A is in $\mathfrak{M}$, then so is its complement A^c.
(3) If A_n is in $\mathfrak{M}$ for $n = 1, 2, 3, \ldots$, then so is $\cup_{n=1}^{\infty} A_n$.

Furthermore, the set function μ must be *countably additive*: If $\{A_n\}$ is a countable collection of pairwise disjoint sets in $\mathfrak{M}$, then

---

[1] This analogy was proposed by Henri Lebesgue himself in a 1926 address in Copenhagen in which he discussed the origins of his ideas for his theory of integration. An English translation of this address can be found in [7] or [29].

$$\mu(\bigcup_{n=1}^{\infty} A_n) = \sum_{n=1}^{\infty} \mu(A_n).$$

To avoid a trivial situation, we also require the existence of some $A \in \mathfrak{M}$ with $\mu(A) < \infty$.

The sets in $\mathfrak{M}$ are called the $(\mu\text{-})$measurable sets. Conditions (2) and (3) say that the collection of measurable sets is closed under complementation and countable unions. The set function μ, whose domain is the collection of all measurable sets in X, is called a *measure* on X. When the σ-algebra $\mathfrak{M}$ or the measure μ is clear from the context, they are often omitted from the notation $(X, \mathfrak{M}, \mu)$. There are such things as signed measures (taking values in the real line $\mathbb{R}$) and complex measures (taking values in $\mathbb{C}$), but unless we say explicitly otherwise, "measure" will always mean "positive measure" in the sense of Definition A.1. If the values of μ are restricted to $[0, \infty)$ it is called a *finite measure*. A measure is said to be σ-*finite* if the underlying set X can be written as a countable union of (measurable) sets each having finite measure.

One can easily obtain the following as consequences of Definition A.1 and simple set-theoretic manipulations:

(a) $\emptyset \in \mathfrak{M}$, and hence a *finite* union of measurable sets is measurable.
(b)   A finite or countable intersection of measurable sets is measurable.
(c) $\mu(\emptyset) = 0$.
(d)   If A and B are measurable sets, with $A \subseteq B$, then $\mu(A) \leq \mu(B)$; this says μ is *monotone*.

We'll see shortly why we want the flexibility to have $\mathfrak{M}$ be a proper subset of the collection $\mathscr{P}(X)$ of all subsets of X. Nevertheless, there are important examples of measure spaces where $\mathfrak{M} = \mathscr{P}(X)$, and hence where the requirements (1)–(3) of Definition A.1 are automatically satisfied.

Example A.2. Let $X = \mathbb{N}$, the natural numbers, set $\mathfrak{M} = \mathscr{P}(\mathbb{N})$, and let μ assign to each finite subset of $\mathbb{N}$ its cardinality, and to each infinite subset of $\mathbb{N}$ the value ∞. With the convention that $a + \infty = \infty + a = \infty$ for $0 \leq a \leq \infty$, verification that μ is countably additive is immediate. This is called *counting measure* on the positive integers. Notice that the only set with counting measure zero is the empty set. Counting measure on $\mathbb{N}$ is not a finite measure, but it is σ-finite.

Example A.3. Let X be any set, let $\mathfrak{M} = \mathscr{P}(X)$ and fix an arbitrary point x_0 in X. Define

$$\mu(A) = \begin{cases} 1 & \text{if } x_0 \in A \\ 0 & \text{otherwise} \end{cases} \tag{A.1}$$

for each $A \subseteq X$. Verification that $(X, \mathfrak{M}, \mu)$ is a measure space is easy. The measure μ is called the (unit) point mass measure at x_0.

Our most important example will be "Lebesgue measure" on the real line $\mathbb{R}$ or on an interval $[a, b] \subseteq \mathbb{R}$. The underlying idea is to generalize the notion of "length" from intervals to more general sets. That is, we seek a measure space $(\mathbb{R}, \mathfrak{M}, m)$ where $m(I) = |I| = d - c$ whenever I is an interval with endpoints c and d. Furthermore, we want this measure m to be *translation invariant*, so that $m(A + x) = m(A)$ for every $A \in \mathfrak{M}$, where

$$A + x \equiv \{a + x : a \in A\}$$

is the translate of A by $x \in \mathbb{R}$.

Perhaps unexpectedly—and this explains why our definition of a measure space doesn't require that $\mathfrak{M} = \mathscr{P}(X)$—it is impossible to do this if we want *every* subset of $\mathbb{R}$ to be measurable. As soon as we ask that the measure of an interval be its length, and require the measure to be translation invariant, there must exist non-measurable sets.[2] Fortunately, it is possible to satisfy our desired properties with a σ-algebra $\mathscr{L}$ of subsets of $\mathbb{R}$ that is sufficiently rich to include all open sets in $\mathbb{R}$. In particular, there is a smallest σ-algebra containing all the open sets, called the *Borel σ-algebra*. In addition to all open sets, the Borel σ-algebra contains all closed sets, any countable union of closed sets, any countable intersection of open sets, and so on. The Lebesgue measurable sets, which we discuss next, form a σ-algebra $\mathscr{L}$ which (properly) contains the Borel sets.

While the details of the construction of $\mathscr{L}$ and of Lebesgue measure on $(\mathbb{R}, \mathscr{L})$ will not be given here, it is easy to give an outline as to how to proceed. For more information the reader is referred to [39], for example. Motivated by the desire to have the measure of an interval be its length, we look at all ways of covering an *arbitrary* set $A \subseteq \mathbb{R}$ by a countable collection of open intervals $\{I_n\}$ and define $m^*(A) \in [0, \infty]$ by

$$m^*(A) = \inf \left\{ \sum_{n=1}^{\infty} |I_n| : A \subseteq \bigcup_{n=1}^{\infty} I_n \right\}.$$

This is called the *Lebesgue outer measure* of A; it is defined for all subsets of $\mathbb{R}$ and is translation invariant. It should also be clear the outer measure is monotone, so that $A \subseteq B$ implies that $m^*(A) \leq m^*(B)$. The outer measure of an interval is its length.

Outer measure fails to be countably additive, but there is a proper subset $\mathscr{L}$ of $\mathscr{P}(\mathbb{R})$ which is a σ-algebra containing all open sets, so that the restriction of m^* to $\mathscr{L}$ is countably additive. The subsets of $\mathbb{R}$ that belong to $\mathscr{L}$ are defined to be those sets A satisfying

$$m^*(T) = m^*(T \cap A) + m^*(T \cap A^c)$$

for *all* $T \subseteq \mathbb{R}$, where A^c is the complement of A in $\mathbb{R}$. This definition, which is not Lebesgue's original one, but is rather due to Carathéodory, is perhaps not completely transparent. We use T for "test" set; we are testing the additivity of the outer measure of arbitrary sets against A. Some observations follow easily from Carathéodory's definition. For example, every set with outer measure zero is Lebesgue measurable (i.e., it belongs to $\mathscr{L}$), and a set belongs to $\mathscr{L}$ if and only if its complement does. With some effort, one shows that the measurable sets form a σ-algebra which contains all intervals, and thus all open sets.

---

[2] For an example of a nonmeasurable set, and an interesting discussion of the role of the axiom of choice in its construction, see [4].

When $A \in \mathscr{L}$, we define its Lebesgue measure $m(A)$ by

$$m(A) = m^*(A);$$

that is, Lebesgue measure is simply the restriction of outer measure to the Lebesgue measurable sets $\mathscr{L}$. Once we have defined Lebesgue measure on $\mathbb{R}$, we can also consider Lebesgue measure on any interval $[a,b]$, by intersecting measurable sets in $\mathbb{R}$ with $[a,b]$. We'll reserve the notation m for Lebesgue measure. Sets of measure zero are easy to understand: A subset of $\mathbb{R}$ has measure zero precisely when for each positive ε it can be covered by an at most countable collection of intervals the sum of whose lengths is less than ε. Countable sets have measure zero, but there are uncountable sets of measure zero as well.[3]

A.2 Integration

Let's recall how the Riemann integral of a bounded, real-valued function on the interval $[a,b]$ is defined. Partition the interval into subintervals by means of subdivision points $a = x_0 < x_1 < x_2 < \cdots < x_n = b$. For each such partition, we have the upper and lower sums

$$U_f = \sum_{j=1}^{n} M_j(x_j - x_{j-1}) \text{ and } L_f = \sum_{j=1}^{n} m_j(x_j - x_{j-1})$$

where

$$M_j = \sup_{x_{j-1} \leq x \leq x_j} f(x) \quad \text{and} \quad m_j = \inf_{x_{j-1} \leq x \leq x_j} f(x).$$

We say that f is Riemann integrable on $[a,b]$ if

$$\inf(U_f) = \sup(L_f),$$

where the infimum and supremum are taken over all partitions of $[a,b]$ as just described. The common value of this infimum and supremum is the Riemann integral of f over $[a,b]$. In the 1820s Cauchy, who is credited with the first attempt at a rigorous definition of continuity, had considered sums of a similar sort (choosing $f(x_{j-1})$ instead of m_j or M_j), but he assumed a priori that f was continuous. By contrast, Riemann as part of the work for his Habilitation degree in 1854, did not suppose f to be continuous, and thus called attention to the question: What functions are (Riemann) integrable? To illustrate some of the nuances of this question, Riemann gave an example of a function whose discontinuities are dense in the the real line, but which is nevertheless Riemann integrable on any finite interval $[a,b]$. It is easy to see, however, that "too many" discontinuities can cause trouble. The example,

---

[3] The Cantor ternary set provides an example, since its complement in $[0,1]$ is a collection of disjoint intervals whose lengths sum to 1.

due to Dirichlet, of the function defined on $[0,1]$ by

$$f(x) = \begin{cases} 1 \text{ if } x \text{ is irrational} \\ 0 \text{ if } x \text{ is rational} \end{cases} \tag{A.2}$$

has $U_f = 1$ and $L_f = 0$ for each partition of $[0,1]$, and hence f is not Riemann integrable. This function is discontinuous at every point. One can show that a bounded function on $[a,b]$ is Riemann integrable if and only if the set of points at which it fails to be continuous has Lebesgue measure zero.[4]

By the 1870s Riemann's theory of integration had become widely known, and had had successful application in a number of areas. Some limitations of Riemann's method had also come to light, but these were not yet regarded as serious deficiencies.

From a modern perspective we can see several problems with Riemann's theory of integration. Most simply stated, not enough functions are Riemann integrable. There is an incompatibility of Riemann integration and limit processes. Every potential difficulty with the statement

$$\lim_{n\to\infty} \int f_n = \int \left(\lim_{n\to\infty} f_n\right), \tag{A.3}$$

where the integrals are Riemann integrals, can occur. For example, one side of Equation (A.3) may fail to exist, even if the other side is perfectly well-behaved, or both may exist, but they fail to agree. While the Lebesgue integral doesn't remove all problems with this interchange of limit and integral, we can give several useful conditions under which Equation (A.3) holds; see Theorems A.4 and A.6 below.

From a functional analysis perspective, there is another serious deficiency of the Riemann integral. The set of all Riemann integrable functions $f : [0,1] \to \mathbb{R}$ in the metric

$$d(f,g) = \int_0^1 |f(x) - g(x)|dx$$

fails to be complete.[5] In other words, in this important metric, there are Cauchy sequences of Riemann integrable functions that fail to converge to a Riemann integrable limit. The Lebesgue integral, introduced by Lebesgue in his doctoral thesis of 1902, led to a resolution of this fundamental problem.

We now turn to the definition of the Lebesgue integral. Because it does not rely on a partitioning of the domain, the definition can be just as easily made for an arbitrary measure space $(X, \mathfrak{M}, \mu)$ as for the particular example $(\mathbb{R}, \mathcal{L}, m)$, and we will do so.

---

[4] A confusion between the measure-theoretic notion of smallness (Lebesgue measure zero) and the topological notion of smallness (nowhere dense, in the language of Section 3.2) muddied some of the initial study of Riemann's notion of integral. There are variants of the Cantor set which are nowhere dense but have positive Lebesgue measure.

[5] Strictly speaking, d is not a metric, since $d(f,g) = 0$ does not imply that $f(x) = g(x)$ at every $x \in X$. This problem can be easily rectified, though; see the discussion in Section A.3 below.

Consider first a function s defined on X and taking only finitely many distinct, nonnegative values $\alpha_1, \alpha_2, \ldots, \alpha_n$. If for each $j = 1, 2 \ldots, n$, the set

$$A_j = s^{-1}(\alpha_j)$$

is in $\mathfrak{M}$, we call s a *nonnegative measurable simple function* (a simple function in general is one taking only finitely many distinct values). The sum

$$\sum_{j=1}^{n} \alpha_j \mu(A_j)$$

is a value in $[0, \infty]$ (we define $\alpha_j \mu(A_j) = 0$ if $\alpha_j = 0$ and $\mu(A_j) = \infty$). We call this the Lebesgue integral of the nonnegative simple function s with respect to μ, and denote it

$$\int_X s \, d\mu.$$

Furthermore, for any measurable subset E of X, define

$$\int_E s \, d\mu = \sum_{j=1}^{n} \alpha_j \mu(A_j \cap E).$$

Notice how we need each A_j to be μ-measurable for these definitions to make sense. An arbitrary real-valued function f on X is said to be *measurable* if

$$f^{-1}[\alpha, \beta) \equiv \{x \in X : \alpha \le f(x) < \beta\}$$

is in $\mathfrak{M}$ for each $\alpha, \beta \in \mathbb{R}$. There is a fair amount of flexibility in this definition. For example, the half-open intervals $[\alpha, \beta)$ can be replaced by open intervals, or closed intervals, or arbitrary open sets, or arbitrary closed sets in $\mathbb{R}$. In measure spaces where all sets are measurable (like that of Example A.2), all functions are measurable.

The definition of measurability shows that if we take a bounded, nonnegative, measurable function f on X, we can approximate f by *measurable* simple functions in the following natural way. Suppose that $0 \le f \le M$ on X and let n be a positive integer. As in the quote of Lebesgue which introduces this chapter, we partition $[0, M]$ into nonoverlapping subintervals I_j by

$$I_1 = \left[0, \frac{1}{n}\right), \qquad I_2 = \left[\frac{1}{n}, \frac{2}{n}\right)$$

and in general

$$I_j = \left[\frac{j-1}{n}, \frac{j}{n}\right)$$

for $j \le Mn + 1$. Define s_n on X to be $\frac{j-1}{n}$ on $f^{-1}(I_j)$ for $j = 1, 2, \ldots, Mn + 1$, so that s_n is a measurable simple function and

$$0 \le f(x) - s_n(x) \le \frac{1}{n}$$

on X.

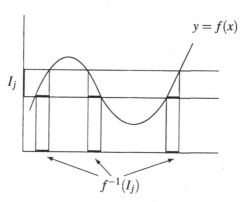

FIGURE A.1: Constructing the measurable simple function s_n

With this approximation scheme to aid our intuition, we define the Lebesgue integral of any measurable $f : X \to [0,\infty]$ as

$$\int_X f \, d\mu = \sup \left\{ \int_X s \, d\mu : s \text{ is a simple measurable function and } 0 \le s \le f \right\},$$

and say that f is Lebesgue integrable if this supremum is finite. We integrate over a measurable subset E of X by defining

$$\int_E f \, d\mu = \sup \left\{ \int_E s \, d\mu : s \text{ is a simple measurable function and } 0 \le s \le f \right\}.$$

Two monotonicity properties follow readily from these definitions:

$$0 \le f \le g \Longrightarrow \int_A f \, d\mu \le \int_A g \, d\mu$$

and

$$0 \le f \text{ and } A \subseteq B \Longrightarrow \int_A f \, d\mu \le \int_B f \, d\mu.$$

In a general measure space $(X, \mathfrak{M}, \mu)$ we can always assume that every subset of any set of μ-measure zero belongs to $\mathfrak{M}$ (see Theorem 1.36 in [40]; we have already observed this property for m). This implies that if a measurable function on X is changed on a set of μ-measure zero, the result is still measurable. Moreover, integrals are not affected by such a change. Thus many results in the theory of Lebesgue integration are stated with the provision "almost everywhere," meaning, except possibly on a set of measure zero. For an example, see Theorem A.6 below.

As a simple exercise in using this terminology, the reader is invited to show that for a *nonnegative* measurable function f and $A \in \mathfrak{M}$,

$$\int_A f \, d\mu = 0 \text{ if and only if } f = 0 \text{ almost everywhere on } A.$$

Lebesgue conceived of his theory of integration as an *extension* of Riemann's, and a Riemann integrable function on an interval $[a, b]$ in $\mathbb{R}$ will also be Lebesgue integrable, with equality of the integrals. The function in (A.2) provides an example of a measurable simple function on $[0, 1]$ with Lebesgue integral equal to 1 that is not Riemann integrable.

We state two theorems about interchange of limit and integral. Both concern sequences of measurable functions on an arbitrary positive measure space $(X, \mathfrak{M}, \mu)$ and their Lebesgue integrals, and give useful conditions under which a statement in the form of Equation (A.3) holds.

Theorem A.4 (Monotone Convergence Theorem). *Suppose that $\{f_n\}$ is a sequence of measurable functions with*

$$0 \leq f_1(x) \leq f_2(x) \leq \cdots \leq \infty$$

for all $x \in X$. If $f(x) = \lim_{n \to \infty} f_n(x)$ for each $x \in X$, then f is measurable, and

$$\lim_{n \to \infty} \int_X f_n \, d\mu = \int_X f \, d\mu.$$

Notice that monotonicity implies that the limit function f always exists, so long as we permit it to take values in $[0, \infty]$.

As an application of Theorem A.4, we encourage the reader to verify the details of the following example.

Example A.5. When μ is counting measure on the set $\mathbb{N}$ of natural numbers,

$$\int_{\mathbb{N}} f \, d\mu = \sum_{n=1}^{\infty} f(n)$$

for any *nonnegative*-valued function f on $\mathbb{N}$. Any such function can be realized in a natural way as a monotone increasing limit of simple functions.

Our next theorem on the interchange of limit and integral is normally stated for complex-valued functions. Before giving its statement, we need to extend our definition of the Lebesgue integral to this larger class. Any measurable real-valued function f can be written as a difference $f^+ - f_-$ of two nonnegative measurable functions, where $f^+ = \max(f, 0)$ and $f_- = -\min(0, f)$. For complex-valued functions $f = u + iv$, measurability is defined by requiring that the real-valued functions u and v be measurable. When f is measurable, so is its modulus $|f|$. This means that the definition of Lebesgue integration can be readily extended to certain complex-valued functions as follows: Provided $\int_X |f| \, d\mu < \infty$, define for any μ-measurable set A in $\mathfrak{M}$

$$\int_A f d\mu = \int_A u^+ d\mu - \int_A u^- d\mu + i \int_A v^+ d\mu - i \int_A v^- d\mu \qquad (A.4)$$

where $f = u + iv$ and $u^+, u^-, v^+,$ and v^- are the positive and negative parts of u and v, respectively. The restriction

$$\int_X |f| d\mu < \infty \qquad (A.5)$$

guarantees that each of the four integrals on the right-hand side of (A.4) is finite. Measurable complex-valued functions which satisfy (A.5) are said to be Lebesgue integrable with respect to μ. Thus the notation $\int_X f \, d\mu$ is used for nonnegative measurable functions, where $\int_X f \, d\mu = \infty$ is possible, and for measurable, complex-valued functions satisfying $\int_X |f| d\mu < \infty$ where $\int_X f \, d\mu$ will be a (finite) value in $\mathbb{C}$. Lebesgue integration is linear, meaning

$$\int_X (\alpha f + g) \, d\mu = \alpha \int_X f \, d\mu + \int_X g \, d\mu$$

for all complex scalars α and complex measurable functions f and g satisfying $\int_X |f| d\mu < \infty$ and $\int_X |g| d\mu < \infty$. While this linearity is crucial, there are some subtleties in proving it.

Theorem A.6 (Lebesgue Dominated Convergence Theorem). *Suppose* $\{f_n\}$ *is a sequence of complex-valued μ-measurable functions and suppose*

$$\lim_{n \to \infty} f_n(x) = f(x)$$

for almost every $x \in X$. If there exists a measurable function g on X with

$$|f_n(x)| \le |g(x)|$$

for almost every $x \in X$ and

$$\int_X |g| d\mu < \infty$$

then f is measurable and

$$\lim_{n \to \infty} \int_X f_n d\mu = \int_X f d\mu.$$

As applications of the convergence theorems A.4 and A.6, and the linearity of the Lebesgue integral, we can give conditions under which term-by-term integration is allowed. This is an historically important issue, and with the Riemann integral instead of the Lebesgue integral sufficient conditions for such integration of series are often too restrictive.

Theorem A.7. *Suppose* $\{f_n\}$ *is a sequence of μ-measurable functions.*

(a) If for all n, $f_n : X \to [0, \infty]$, and we define

$$f(x) = \sum_{n=1}^{\infty} f_n(x),$$

then

$$\int_X f \, d\mu = \sum_{n=1}^{\infty} \int_X f_n \, d\mu.$$

(b)   *If the f_n are complex-valued and*

$$\sum_{n=1}^{\infty} \int_X |f_n| \, d\mu < \infty,$$

then the series $\sum_{n=1}^{\infty} f_n$ converges almost everywhere, and

$$\int_X \left(\sum_{n=1}^{\infty} f_n \right) d\mu = \sum_{n=1}^{\infty} \int_X f_n \, d\mu.$$

Part (a) is proved by applying the monotone convergence theorem to the sequence of partial sums of $\sum_{n=1}^{\infty} f_n$. We allow the possibility that the positive term series $\sum_{n=1}^{\infty} f_n(x)$ fails to converge for some x, in which case $f(x) = \infty$. Part (b) is proved by applying the dominated convergence theorem to the sequence of partial sums, which are dominated by $g(x) \equiv \sum_{n=1}^{\infty} |f_n(x)|$, where $\int_X g \, d\mu < \infty$ by (a).

A.3 L^p Spaces

For any measure space $(X, \mathfrak{M}, \mu)$ and $1 \leq p < \infty$, set

$$L^p(X, \mu) \equiv \{\text{complex-valued measurable } f : \int_X |f|^p d\mu < \infty\}.$$

The notation $L^p(X, \mu)$ is often shortened to $L^p(X)$ or $L^p(\mu)$ when no confusion can result. We also define the space $L^\infty(X, \mu)$ of *essentially bounded* functions. We say that a measurable function is essentially bounded if there exists $M < \infty$ so that

$$\mu(\{x : |f(x)| > M\}) = 0. \tag{A.6}$$

Equivalently, f is essentially bounded if it can be changed on a set of measure zero to produce a bounded function.

When $1 \leq p < \infty$, $L^p(X, \mu)$ is a normed linear space if we define

$$\|f\|_p \equiv \left(\int_X |f|^p d\mu \right)^{\frac{1}{p}},$$

provided we agree to identify functions which agree μ-almost everywhere. If $f \in L^\infty(X, \mu)$, the infimum of all $M > 0$ that satisfy (A.6) is called the essential supremum of f, and is denoted $\|f\|_\infty$. This definition makes $L^\infty(X,\mu)$ a normed linear space.

To verify the assertions of the last paragraph there are two issues to check: Is the sum of two L^p functions still in L^p, and does the triangle inequality hold for $\|\cdot\|_p$? For $p=1$ and $p=\infty$, these hold as immediate consequences of the triangle inequality $|f+g| \le |f|+|g|$ on $\mathbb{C}$. When $1 < p < \infty$ we need Minkowski's inequality: For measurable functions f and g on X,

$$\left(\int_X |f+g|^p d\mu\right)^{\frac{1}{p}} \le \left(\int_X |f|^p d\mu\right)^{\frac{1}{p}} + \left(\int_X |g|^p d\mu\right)^{\frac{1}{p}}.$$

Minkowski's inequality follows from Hölder's inequality

$$\int_X |fg| d\mu \le \left(\int_X |f|^p d\mu\right)^{\frac{1}{p}} \left(\int_X |g|^q d\mu\right)^{\frac{1}{q}} \tag{A.7}$$

where $1/p + 1/q = 1$ (that is, p and q are *conjugate indices*.) Hölder's inequality holds for the pair $p=1, q=\infty$ if we replace the second integral on the right-hand side of (A.7) by $\|g\|_\infty$. For the proofs of these two basic inequalities see Theorem 3.5 in [40], for example.

By virtue of Example A.5, the space ℓ^p, as defined in Example 1.5 of Chapter 1, is the same as $L^p(\mathbb{N},\mu)$, where μ is counting measure on the subsets of $\mathbb{N}$. This allows us to subsume the theory of ℓ^p into the theory of L^p for general measure spaces. Since the only set with counting measure zero is the empty set, no "almost everywhere" conventions are needed with ℓ^p.

The next result is fundamental. It asserts the completeness of the metric space $L^p(X,\mu)$.

Theorem A.8 (Riesz–Fischer Theorem). *For every positive measure μ and $1 \le p \le \infty$, $L^p(\mu)$ is a Banach space.*

As discussed in Chapter 1, this theorem goes by the name of the Riesz–Fischer theorem, for simultaneous and independent work of Riesz and Fischer in the case $p=2$. Fischer explicitly noted that Theorem A.8 requires "the use of notions of M. Lebesgue," and that completeness does not hold if one considers continuous functions on $[a,b]$ in the $L^2(m)$ metric, since the L^2-limit of a sequence of continuous functions need not be continuous.

A.4 The Stone–Weierstrass Theorem

Recall that for X a compact Hausdorff space, $C(X)$ denotes the continuous complex-valued functions on X, endowed with the supremum norm

$$\|f\|_\infty = \sup\{|f(x)| : x \in X\}.$$

The real-valued functions in $C(X)$ are denoted $C_{\mathbb{R}}(X)$; this is a real vector space.

The classical Weierstrass theorem says that for any finite interval $[a,b]$ of the real line, the (real-valued) polynomials are dense in $C_{\mathbb{R}}[a,b]$; that is, given any $f \in C_{\mathbb{R}}[a,b]$, there exist polynomials P_n converging uniformly to f on $[a,b]$. The same result holds for $C[a,b]$, except now the polynomials are allowed to be complex-valued. This is a fundamental result in analysis, and it has numerous proofs.

When the interval $[a,b]$ is replaced by a compact Hausdorff space, it is not immediately clear how one might generalize Weierstrass's result, as there is no notion of polynomials on general spaces. However, since the (real) polynomials on an interval are generated from sums, products, and real scalar products of the functions $f(x) = 1$ and $g(x) = x$, this suggests consideration of subalgebras of $C(X)$.

Recall that a (closed) subalgebra $\mathscr{B}$ of $C(X)$ (or $C_{\mathbb{R}}(X)$) is a (closed) subspace that is closed under multiplication. We say that a subalgebra of $C(X)$ separates points if given any $x,y \in X$ there is an $f \in \mathscr{B}$ with $f(x) \neq f(y)$. Marshall Stone generalized the Weierstrass theorem as follows: If $\mathscr{B}$ is a closed subalgebra of $C_{\mathbb{R}}(X)$ that separates points and contains the constant functions, then $\mathscr{B} = C_{\mathbb{R}}(X)$. This is sometimes called the real Stone–Weierstrass theorem; a complexified version requires one additional hypothesis, namely that $\mathscr{B}$ be closed under conjugation (meaning that $f \in \mathscr{B}$ implies $\bar{f} \in \mathscr{B}$).

Theorem A.9 (Stone–Weierstrass Theorem). *If $\mathscr{B}$ is a closed subalgebra of $C(X)$ that separates points, contains the constant functions, and is closed under conjugation, then $\mathscr{B} = C(X)$.*

To see why the extra hypothesis is needed, take X to be the closed unit disk $\overline{\mathbb{D}}$ in $\mathbb{C}$, and let $\mathscr{B}$ be the closed subalgebra of functions that are continuous in the closed disk and analytic in its interior, so that $\mathscr{B}$ separates points, contains the constants, but is not all of $C(\overline{\mathbb{D}})$, since, for example, $f(z) = \bar{z}$ does not belong to $\mathscr{B}$.

A proof of the Stone–Weierstrass Theorem can be found in [36].

A.5 Positive Linear Functionals on $C(X)$

Let X be a compact Hausdorff space. The reader may find it convenient to think of X as a compact subset of the complex plane, since our principle application is to this setting.

A positive linear functional on $C(X)$ is a linear functional $\Lambda : C(X) \to \mathbb{C}$ with the property that $\Lambda(f) \geq 0$ if $f \geq 0$ on X. We don't a priori assume that Λ is bounded, but as a consequence of the positivity hypothesis, it will be so.

By the Borel sets in X, we mean the smallest σ-algebra that contains all open sets in X. A measure defined on this σ-algebra is called a *Borel measure*, and a function which is measurable with respect to the Borel σ-algebra is called a *Borel measurable function*, or simply a Borel function. It should be clear that any function in $C(X)$ is a Borel function. If we start with a finite, positive, Borel measure on X then $C(X) \subseteq L^1(X, \mu)$, and the mapping

$$f \rightarrow \int_X f \, d\mu$$

will be a positive linear functional on $C(X)$. The following result, known as the Riesz–Markov theorem, says that all positive linear functionals arise in this way.

Theorem A.10 (Riesz–Markov Theorem). *If X is a compact Hausdorff space and Λ is a positive linear functional on $C(X)$, then there is a unique (positive) regular Borel measure μ on X with*

$$\Lambda(f) = \int_X f \, d\mu$$

for all $f \in C(X)$.

The adjective "regular" which appears in the statement will not be defined precisely here; see, for example [40]. Borel measures that are *not* regular are rather pathological. When X is a compact subset of the complex plane, for example, all Borel measures on X are regular. Notice that the norm of the linear functional Λ is $\mu(X)$.

References

1. D. Amir, *Characterizations of Inner Product Spaces*, Operator Theory: Advances and Applications, Vol. 20, Birkhäuser, Basel, 1986.
2. W. Arveson, *A Short Course on Spectral Theory*, Springer-Verlag, New York, 2002.
3. B. Beauzamy, *Introduction to Operator Theory and Invariant Subspaces*, North-Holland, Amsterdam, 1988.
4. D. Bressoud, *A Radical Approach to Lebesgue's Theory of Integration*, MAA Textbooks, Cambridge University Press, Cambridge, 2008.
5. S. Brown, Some invariant subspaces for subnormal operators, *Integral Equations Operator Theory* 1 (1978), 310–333.
6. N. Bourbaki, *Eléments de Mathématique: Espaces Vectoriels Topologiques Chap. I–V*, Masson, Paris, 1981.
7. S.B. Chae, *Lebesgue Integration*, Second Edition, Springer-Verlag, New York, 1995.
8. J. Conway, *A Course in Functional Analysis*, Second Edition, Springer-Verlag, New York, 1990.
9. A. M. Davie, Review of "On the invariant subspace problem for Banach spaces" by P. Enflo, Mathematical Reviews on the Web, MR892591, 1988. http://www.ams.org/mathscinet
10. J. Dieudonné, *History of Functional Analysis*, Mathematics Studies 49, North-Holland, Amsterdam, 1981.
11. R. Douglas, *Banach Algebra Techniques in Operator Theory*, Pure and Applied Mathematics, Vol. 49. Academic Press, New York–London, 1972.
12. N. Dunford and J. Schwartz, *Linear Operators, Part I: General Theory*, Interscience, New York, 1958.
13. M. Fréchet, *Espaces abstraits*, Gauthier-Villars, Paris, 1928.
14. C. Goffman and G. Pedrick, *First Course in Functional Analysis*, Prentice Hall, Englewood Cliffs, 1965.
15. P. Halmos, What does the spectral theorem say? *Amer. Math. Monthly* 70 (1963), 241–247.
16. P. Halmos, The legend of John von Neumann, *Amer. Math. Monthly* 80 (1973), 382–394.
17. P. Halmos, *A Hilbert Space Problem Book*, Second Edition, Springer-Verlag, New York, 1982.
18. P. Halmos, *I Want to be a Mathematician: An Automathography*, Springer-Verlag, New York, 1985.
19. D. Hilbert, *Grundzüge einer allgemeinen Theorie der linearen Integralgleichungen*, Chelsea, New York, 1953.
20. H. Hochstadt, Eduard Helly, father of the Hahn-Banach theorem, *Math. Intelligencer* 2 (1979/1980), No. 3, 123–125.
21. J. Horváth, On the Riesz–Fischer theorem, *Studia Sci. Math. Hungar.* 41 (2004), 467–478.
22. E. Helly, Über Systeme linearer Gleichungen mit unendlich vielen Unbekannten, *Monatsh. Math. Phys.* 31 (1921), 60–91.

23. R.C. James, A non-reflexive Banach space isometric with its second conjugate space , *Proc. Nat. Acad. Sci. U.S.A.* **37** (1951), 174–177.

24. R.C. James, Characterizations of reflexivity, *Studia Math.* **23** (1963/1964), 205–216.

25. R. Kadison, Notes on the Gelfand–Neumark Theorem, in C^*-Algebras: 1943–1993, A Fifty Year Celebration, Contemp. Math., **167**, Amer. Math. Soc., Providence, RI 1994.

26. R. Kaluza, *Through a reporter's eyes: The Life of Stefan Banach*, Birkhäuser, Boston, 1996.

27. S. Krantz, *Mathematical Aprochypha*, Mathematical Association of America, Washington DC, 2002.

28. S. Krein and Yu. Petunin, Scales of Banach spaces, *Russian Math. Surveys* **21** (1966), 85–159.

29. H. Lebesgue, *Measure and the Integral*, Holden-Day, San Francisco, 1966.

30. N. Macrae, *John von Neumann*, Amer. Math. Soc., Providence, 1999.

31. R.D. Mauldin (Ed.), *The Scottish Book: Mathematics from the Scottish Café*, Birkhäuser, Boston, 1981.

32. V. Milman, Observations on the movement of people and ideas in twentieth-century mathematics, 215–242, in *Mathematical Events of the Twentieth Century*, A. Bolibruch, Yu. Osipov, YA. Sinai (Eds.), Springer-Verlag, Berlin, PHASIS, Moscow, 2006.

33. J. Munkres, *Topology: A First Course*, Prentice-Hall, Englewood Cliffs, NJ, 1975.

34. A. Pietsch, *History of Banach Spaces and Linear Operators*, Birkhäuser, Boston, 2007.

35. H. Radjavi and P. Rosenthal, *Invariant Subspaces*, Second Edition, Dover Publications, Mineola, NY, 2003.

36. M. Reed and B. Simon, *Methods of Modern Mathematical Physics I: Functional Analysis*, Second Edition, Academic Press, New York, 1980.

37. M. Rédei, Ed., *John von Neumann: Selected Letters*, Amer. Math. Soc., London Math., Soc., Providence, 2005.

38. C. Reid, *Hilbert*, Springer-Verlag, Berlin, 1970.

39. H. Royden, *Real Analysis*, Third Edition, Macmillan, New York, 1988.

40. W. Rudin, *Real and Complex Analysis*, Third Edition, McGraw-Hill, New York, 1987.

41. D. Sarason, The multiplication theorem for Fredholm operators, *Amer. Math. Monthly* **94** (1987), 68–70.

42. D. Sarason, The exact answer to a question of Shields, *Math. Intelligencer* **12**, (1990), 18–19.

43. K. Saxe, *Beginning Functional Analysis*, Springer, New York, 2002.

44. H. Steinhaus, Stefan Banach, *Studia Math. (Ser. Specjalna) Zesyt* **1** (1963), 7–15.

45. A. Vershik, The life and fate of functional analysis in the twentieth century, 437–447, in *Mathematical Events of the Twentieth Century*, A. Bolibruch, Yu. Osipov, YA. Sinai (Eds.), Springer-Verlag, Berlin, PHASIS, Moscow, 2006.

46. N. Wiener, *I Am a Mathematician*, Doubleday, New York, 1956.

47. L. Young, *Mathematicians and Their Times*, Mathematics Studies 48, North-Holland, Amsterdam, 1981.

48. N. Young, *An Introduction to Hilbert Space*, Cambridge University Press, Cambridge, 1988.

Index